Antibiotics II

Japanese Technology Reviews

Section E: Biotechnology

Volume 1, Number 1
Production of Nucleotides and Nucleosides by Fermentation
Sadao Teshiba and Akira Furuya

Volume 1, Number 2
Recent Progress in Microbial Production of Amino Acids
Hitoshi Enei, Kenzo Yokozeki and Kunihiko Akashi

Volume 2, Number 1
Recent Advances in Japanese Brewing Technology
Takashi Inoue, Jun-ichi Tanaka and Shunsuke Mitsui

Volume 2, Number 2
Antibiotics I: β-Lactams and Other Antimicrobial Agents
Isao Kawamoto and Masao Miyauchi

Volume 3, Number 1
Antibiotics II: Antibiotics by Fermentation
Sadao Teshiba, Mamoru Hasegawa, Takashi Suzuki, Yoshiharu Tsubota, Hidehi Takebe, Hideo Tanaka, Mitsuyasu Okabe and Rokuro Okamoto

Volume 3, Number 2
Oligosaccharides: Production, Properties, and Applications
Edited by Teruo Nakakuki

Antibiotics II

Antibiotics
by Fermentation

Sadao Teshiba, Mamoru Hasegawa,
Takashi Suzuki, Yoshiharu Tsubota,
Hidehi Takebe, Hideo Tanaka,
Mitsuyasu Okabe and Rokuro Okamoto

CRC Press
Taylor & Francis Group
Boca Raton London New York

CRC Press is an imprint of the
Taylor & Francis Group, an **informa** business
AN A K PETERS BOOK

GENERAL INFORMATION

Aims and Scope

Japanese Technology Reviews is a series of tracts which examines the status and future prospects for Japanese technology.

CRC Press
Taylor & Francis Group
6000 Broken Sound Parkway NW, Suite 300
Boca Raton, FL 33487-2742

First issued in hardback 2019

ISBN-13: 978-2-88124-889-4 (pbk)
ISBN-13: 978-1-138-45718-8 (hbk)

Visit the Taylor & Francis Web site at
http://www.taylorandfrancis.com

and the CRC Press Web site at
http://www.crcpress.com

Distributed by STBS - Publishers Distributor.

April 1993

Contents

4. Novel Acyl Derivatives of Tylosin Produced by Microbial Transformation

Mitsuyasu Okabe and Rokuro Okamoto

Preface to the Series

Modern technology has a great impact on both industry and society. New technology is first created by pioneering work in science. Eventually, a major industry is born, and it grows to have an impact on society in general. International cooperation in science and technology is necessary and desirable as a matter of public policy. As development progresses, international cooperation changes to international competition, and competition further accelerates technological progress.

Japan is in a very competitive position relative to other developed countries in many high technology fields. In some fields, Japan is in a leading position; for example, manufacturing technology and microelectronics, especially semiconductor LSIs and optoelectronic devices. Japanese industries lead in the application of new materials such as composites and fine ceramics, although many of these new materials were first developed in the United States and Europe. The United States, Europe and Japan are working intensively, both competitively and cooperatively, on the research and development of high-critical-temperature superconductors. Computers and communications are now a combined field that plays a key role in the present and future of human society. In the next century, biotechnology will grow, and it may become a major segment of industry. While Japan does not play a major role in all areas of biotechnology, in some areas such as fermentation (the traditional technology for making sake), Japanese research is of primary importance.

Today, tracking Japanese progress in high technology areas is both a necessary and rewarding process. Japanese academic institutions are very active; consequently, their results are published in scientific and technical journals and are presented at numerous meetings where more than 20,000 technical papers are presented orally every year.

However, due principally to the language barrier, the results of academic research in Japan are not well-known overseas. Many in the United States and in Europe are thus surprised by the sudden appearance of Japanese high technology products. The products are admired and enjoyed, but some are astonished at how suddenly these products appear.

With the series Japanese Technology Reviews, we present state-of-the-art Japanese technology in five fields:

Electronics

Computers and Communications

New Materials

Manufacturing Engineering

Biotechnology

Each tract deals with one topic within each of these five fields and reviews both the present status and future prospects of the technology, mainly as seen from the Japanese perspective. Each author is an outstanding scientist or engineer actively engaged in relevant research and development.

The editors are confident that this series will not only give a deep insight into Japanese technology but will also be useful for developing new technology of interest to our readers.

As editor in chief, I would like to sincerely thank the members of the editorial board and the authors for their contributions to this series.

TOSHIAKI IKOMA

Contributors

Mamoru Hasegawa
Kyowa Hakko Kogyo Co. Ltd
Tokyo
Japan

Mitsuyasu Okabe
Shizuoka University
Japan

Rokuro Okamoto
Sanraku Inc.
Kanagawa
Japan

Hideo Tanaka
University of Tsukuba
Japan

Hidehi Takebe
Meiji Seika Kaisha Ltd
Kanagawa
Japan

Sadao Teshiba
Kyowa Hakko Kogyo. Co. Ltd
Tokyo, Japan

Yoshiharu Tsubota
Takeda Chemical Industries Ltd
Osaka
Japan

Takashi Suzuki
Takeda Chemical Industries Ltd
Tokyo, Japan

CHAPTER 1

Astromicin

Sadao Teshiba and Mamoru Hasegawa

1.1. Strain Improvement and Optimization of Culture Conditions

1.1.1. Discovery

There were already over 100 antibiotics commercially available for therapy when we started our screening program for new antibiotics. Opportunistic infection and the appearance of clinical isolates resistant to a variety of antibiotics, however, still remained great problems in therapy. We decided to screen new basic and water-soluble antibiotics, that is, aminoglycoside antibiotics for the following reasons: (1) they exhibit potent and broad spectrum against Gram-positive and negative bacteria, (2) their parental compounds isolated from nature have commonly been used in clinical tests unlike β-lactam antibiotics whose major compounds are semisynthetic, and (3) their producing strains that often accumulate numerous and various minor components never isolated together with major ones.

Rare actinomycetes from an unusual ecological niche, such as soil from hot springs, lakes, and river water, and/or sediments, composts, etc., were applied to our screening, because they provide new antibiotics of potential interest, excluding known and unwanted antibiotics produced by common microorganisms. For the enrichment of the rare actinomycetes from the large and common population of microorganisms in such sources, various methods, such as heat treatment of soil, a baiting method using pine pollen, and the addition of soil extracts to selecting agar plates, were attempted.

Pure cultures of potential and interesting microorganisms were screened on several kinds of media differing in carbon, nitrogen source, and other constituents and buffer for a pH range. To detect a very small amount of a potentially new and interesting antibiotic, we devised a screening method that is sensitive, rapid, inexpensive, and specific for aminoglycoside antibiotics. Cultures were first enriched with an ion-exchange resin, that is, IRC-50, to absorb basic and water-soluble components specifically, and then were analyzed for antibiotic activity after adjusting samples to pH 8.0, because aminoglycoside antibiotics generally show stronger antibacterial activity in the alkaline range than in the neutral range of pH. Knowledge of the spectrum of activities with many specific strains and chromatographic behavior was also invaluable to eliminate known and unwanted antibiotics. Thus, we could detect potentially new and interesting aminoglycoside antibiotics in cultures at a very early stage of screening.

In 1977, *Micromonospora olivasterospora*, which was named later, isolated from soil in Hiroshima, was found to produce several mixtures of new antibiotics[1-3]. This organism was identified as a new species by microscopic characteristics and physiological properties such as its utilization of sugars and nitrogen sources. Colonies are dark yellow to light wheat, turning to olive or dark green with spore formation. There are not aerial mycelium. The spores are sessile or borne singly on short sporophores from substrate mycelia, about 1.0 μm in diameter, ovoid to spherical in shape, and bluntly spiny-surfaced (Figure 1). It is noteworthy that several other strains capable of producing astromicin were isolated from soils widely distributed throughout the world. They were from soils in Hiroshima and Ibaragi prefectures in Japan, and in North Chicago and West Virginia, in the United States.

The most active member of the complex of antibiotics, astromicin (fortimicin A), though similar to gentamicin C2 in chromatographic behavior, was initially detected by the use of bioautography with two test organisms, a sensitive *E. coli* and a gentamicin-resistant *E. coli* possessing an aminoglycoside 2″-nucleotidyltransferase ANT(2″). Astromicin could clearly be differentiated from gentamicin C2 in two bioautograms, because astromicin was active against the resistant *E. coli*, but gentamicin C2 was inactive. Astromicin was further shown to exhibit a potent, unique, and broad spectrum against Gram-positive and-negative bacteria. Astromicin is also very active against all the resistant microorganisms pos-

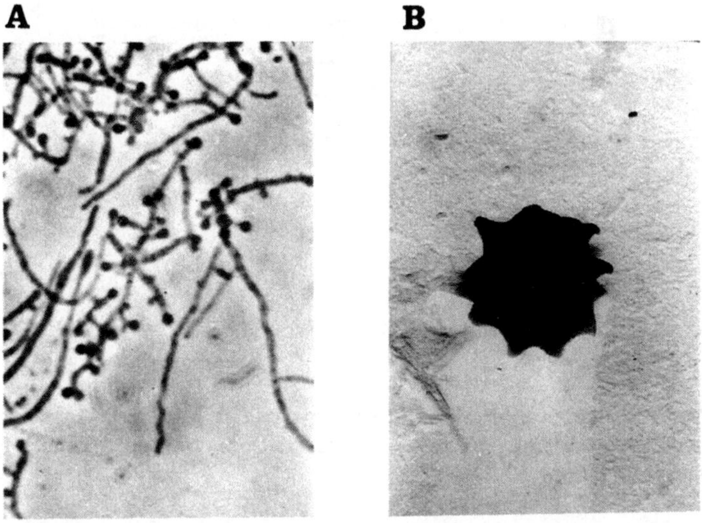

Figure 1. Mycelium (A) and spore (B) of *M. olivasterospora*.

sessing aminogycoside-inactivating enzymes: aminoglycoside 3′-phosphotransferase I and II [APH(3′)-I and APH(3′)-II], aminoglycoside 6′-N-acetyl-transferase [AAC(6′)], aminoglycoside 2′-N-acetyl-transferase [AAC(2′)-II and ANT(2″)]. A sole exception is gentamicin acetyltransferase type I [AAC(3)], which proved to inactivate astromicin by 1-N-acetylation of the fortamine (Figure 2)[4,5].

The structure of astromicin was determined by proton magnetic resonance, carbon magnetic resonance, mass spectra and CD combined with chemical degradation. Astromicin is a pseudodisaccharide and incorporates a novel aminocyclitol, fortamine, in contrast to the diaminocyclitol moieties of known aminoglycosides. Structural characteristics of astromicin explain well the activity against almost all resistant bacteria to aminoglycosides. During the studies, more than 50 minor components[7] were isolated and determined in structure, including secondary major components, such as fortimicin B in which glycine amide moiety of fortamine is removed. Fortimicin B is very weakly active, indicating that the removal of

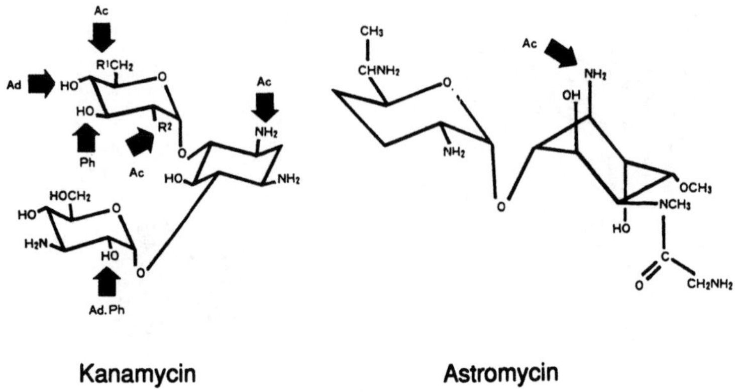

Figure 2. Inactivation sites of kanamycin and astromicin by aminoglycoside-modifying enzymes. Ac: acetylation; Ad: adenylation; Ph: phosphorylation.

glycine from astromicin reduces the activity markedly. After the discovery of astromicin, several related antibiotics were reported from different laboratories (see Chapter 2).

Parallel with strain improvement, clinical tests have been performed in Japanese hospitals and proved the usefulness of astromicin in chemotheray for infectious disease. Figure 3 shows an example of in vitro activity of astromicin to 36 gentamicin-resistant strains of *Serratia marcescence* clinically isolated. While sisomicin, gentamicin, and tobramycin showed weak activity to these strains, astromicin was proved to exhibit more potent activity than semisynthetic aminoglycosides antibiotics such as amikacin, netilmicin, and dibekacin[8,9]. This result is well coincident with the fact that astromicin has only one site (N1 of fortamine) of sugar moieties that can be attacked by inactivating enzymes ad described before. Another great problem of aminoglycoside antibiotics for therapy is the inherited toxicity to auditory organs and the kidney, although they have strong and broad bactericidal activity compared to other types of antibiotics such as β-lactams and macrolides. Astromicin, however, showed remarkably low toxity to auditory organs and the kidney in clinical tests (Figure 4)[10–13]. After several years of clinical tests, astromicin emerged in commercial markets in Japan in 1985

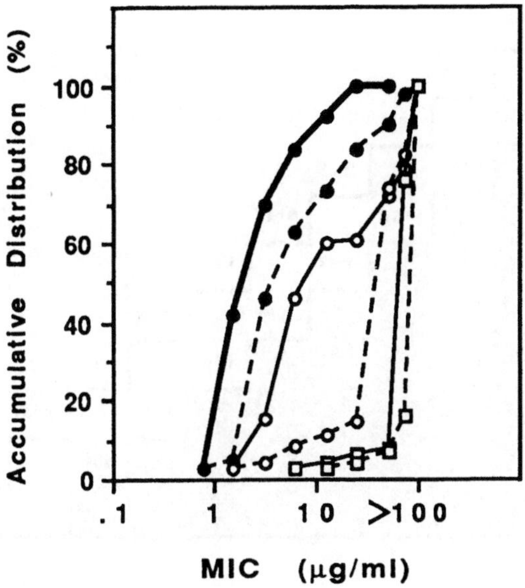

Figure 3. Comparison of antibacterial activity of various aminoglycosides against gentamicin-resistant strains of *S. marcescens*. MIC: minimum inhibitory concentration; (—●—): astromicin; (- - ● - -): amikacin; (—○—): netilmicin; (- - ○ - -): sisomicin; (—□—): tobramycin; (- - □ - -): dibekacin.

and was used for more than half a million patients with various infectious disease in 1987.

1.1.2. Strain Improvement

In view of its broad antibacterial activity in vivo and in vitro and its low acute toxicity, astromicin appeared as a medically useful antibiotic. For further evaluation and industrialization of astromicin, fermentation development such as modifications to the medium, fermentation conditions, and strain improvement have been carried out.

At the earlier stages of strain improvement, spores scraped from sporulation agar plates were mutagenized. Sheared mycelia con-

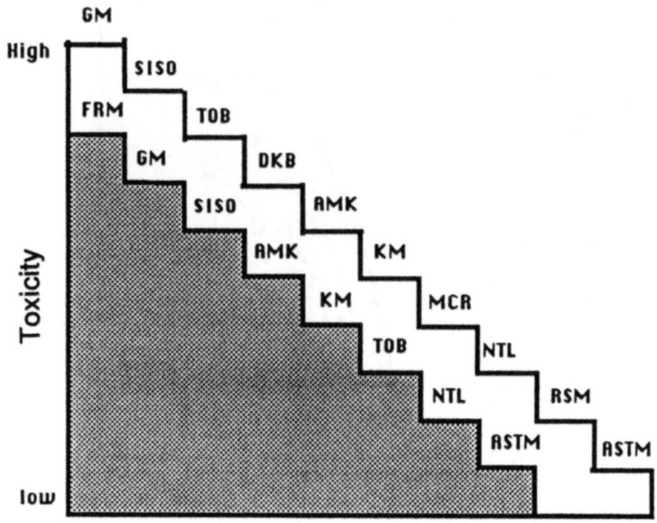

Figure 4. Comparison of renal and aural toxity of various aminoglycosides. Open bars: aural toxicity; shadowed bars: renal toxity; ASTM: astromicin; GM: gentamicin; SISO: sisomicin; TOB: tobramycin; DKB: dibekacin; AMK: amikacin; KM: kanamycin; MCR: micronomicin; NTL: netilmicin; RSM: ribostamycin; FRM: fradiomycin.

sisting of 5 to 10 cells with a mechanical homogenizer were also used for mutational works, because mutants selected at the later stage were poorly sporulated. Protoplasts were also mutagenized to overcome the problem of the lack of a uninuclear stage of filamentous mycelia. Alternatively mutagenized mycelia were converted to protoplasts and then regenerated to colonies (mycelia form). However, protoplast mutation was not so effective in this case. A variety of physical (X-ray and UV irradiation) and chemical mutagens (NTG, EMS, nitrous acid, etc.) was tested.

In parallel with so-called "random" mutation and selection, various "rational" selections were attempted as soon as knowledge of the structure and the biosynthetic pathway was accumulated:

1. Selection of mutants resistant to toxic compounds.

2. Selection of auxotrophic mutants followed by prototrophic reversion.
3. Selection of revertants from a nonproducing strain.
4. Selection of resistant strains to amino acid or sugar analogs.
5. Selection of a self-resistant strain.
6. Selection of morphological mutants including sporulationless and pigmentless strains.
7. Selection after protoplast fusion.

Most attempts were unsuccessful in increasing productivity, but were fruitful in improving the characteristics of producing strains in some cases. Some features of these mutations and selections are described in what follows.

Resistant strains to various antibiotics and uncouplers were isolated and tested for productivity. Among them, some sodium azide-resistant strains showed markedly increased productivity in flask cultures, but failed in large-scale cultures (jar fermentors). A variety of auxotrophic mutants requiring amino acid, nucleobase, or vitamin was attempted to be isolated, but other auxotrophic mutants except methionine and/or cystein, arginine, and uracil could not be isolated. Superior strains compared to the preceding parent strain could not be selected from the auxotrophs and their prototrophic revertants.

In the course of strain improvement, various nonproducing strains have accidentally been obtained. Almost all nonproducing strains were identified as blocked mutants impaired in the biosynthesis of astromicin by cosynthesis or identification of intermediates produced in the medium. Exceptional mutants requiring vitamin B12 for astromicin production, but not for growth, were also isolated as a nonproducing strain. Mutagenized spores or mycelia of nonproducing strains were directly plated after appropriate dilution, and tested for their activity by overlay bioassay using sensitive bacteria.

Proportional correlation between productivity and self-resistance to astromicin has been observed among the producing strains isolated. So we intensively isolated mutants with increased autoresistance, either by adaptation to a gradually increasing concentration of astromicin or by similar selection after mutagenesis. No improved strains in the productivity, however, were selected and nearly all strains showed decreased productivity. Probably only membrane-permeable mutants of astromicin were selected instead of mutants altered in the biosynthetic pathway.

Few reports were available about protoplasting and/or regeneration of rare actinomycete including the *Micromonospora* genus when we started the strain improvement. We developed the optimal conditions of protoplasting and regeneration of *M. olivasterospora*.

Special conditions such as preculture before proplast formation, composition of the regeneration medium, and regeneration conditions were required. Another problem to be overcome was that there were few auxotrophic markers available for selection of fusants. Genetic markers in spore formation, pigment formation, or antibiotics resistance were used instead of an auxotrophic marker. Protoplast fusion between nonproducing strains impaired in the different steps of biosynthetic pathway with each other was also attempted. Recombination frequency was relatively high (more than 10%), and interesting strains were selected with high efficiency from a number of fusants.

In any of the previous cases described, improved mutants often produced only slightly more astromicin (10–15% greater) than the preceding parent strain. How to screen interesting strains efficiently and accurately from a numerous mass is always a key point in a strain-improvement program. Strains selected by agar plate assay with overlaying sensitive bacterium or a test tube culture were successively transferred to a flask culture: 2 liters, 5 liters, 30 liters of jar fermentor tests. Another way to find a strain with slightly increased productivity is the accurate assay of the products accumulated in the medium. The amount of astromicin in the culture fluid was determined by bioassay using sensitive bacterium at the early stage of program, and by thin-layer chromatography (TLC) followed by a densitometry assay at the later stage. The TLC method was very effective, because aminoglycoside antibiotics including astromicin are generally produced with many other minor components. Assay of the minor components in the culture was also helpful for the analysis of the biosynthetic pathway and its regulation. Candidate strains for a following parent for improvement were selected from the viewpoint of practical advantages such as productivity, rate of growth, genetic stability of growth and productivity, and coproduction of minor components.

In summary, the productivity of astromicin was increased a hundred times that of the wild-type strain after strenuous efforts of strain improvement and optimization of culture conditions.

1.1.2.1. Optimization of Culture Conditions

In parallel with the strain-improvement method previously described, optimization of culture conditions including the composition of the media has been carried out with the goal of industrial production.

Optimal conditions naturally depend on the strain investigated. The fundamental composition of the fermentation medium and the effect of carbon sources on astromicin production for use in 5-liter fermentors are shown in Tables 1 and 2, respectively. Among

Table 1. Composition of fermentation medium.

Constituent	Concentration
Maltose	4.5%
Glucose	0.5%
Yeast	4%
Peptone	0.7%
Soybean meal	1.5%
$Mg_3(PO_4)_2$	1%
KH_2PO_4	0.1%
K_2HPO_4	0.1%
$MgSO_4 \cdot 7H_2O$	0.05%
NaCl	0.2%
Ca-D-pantothenate	10 mg/l
Metal mixture	1 ml/l

Table 2. Effect of carbon source.

Carbon source	Productivity
Wheat flour	100
White rice bran	96
Corn starch	79
Maltose	82

various carbon sources examined, natural sources, such as wheat flour or white rice bran, gave the highest yield of astromicin compared to other refined sugars and carbohydrates. However, natural sources with high and homogenous quality were considered relatively difficult to purchase for a long time. Another problem in using these sources that was not easily solved in practical production using large fermentors was their insolubility in aqueous medium. Finally, maltose was chosen as a better carbon source of a fermentation medium. Although glucose could be assimilated more rapidly by producing strain, it did not afford high productivity. Among a variety of nitrogen sources examined, yeast, peptone, and soybean meal (SBM) and their combination accelerated the production of astromicin. Although yeast was absolutely essential for the high productivity, productivity was markedly influenced by the genus, species, and/or strain of yeasts. We had to screen the best strain

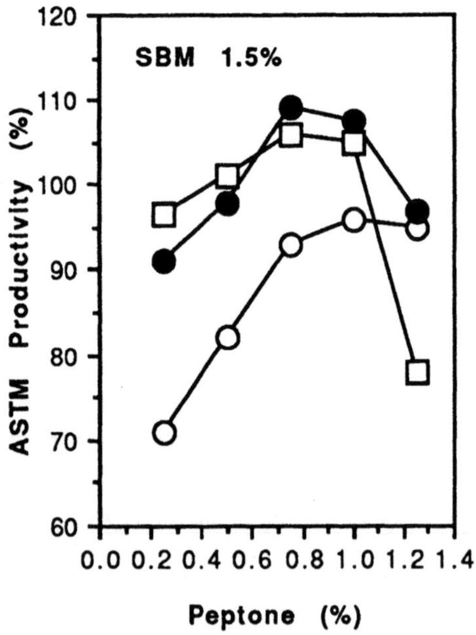

Figure 5. Effect of peptone concentrations on astromicin production in 5-liter jar fermentors in the presence of 4% (○), 5% (●), and 6% (□) yeast.

from a number of strains for its effect, mainly *S. cerevisiae*. Figure 5 shows the effect of peptone and yeast concentrations on astromicin production when used in combination.

The addition of an exchange resin, for example, IRC-50 or CG-50, was significantly effective, not only for the production of astromicin, but also for the growth of producing strains. It increased productivity 30 to 50% both in a flask culture and in jar fermentors. This increase was probably caused by the release of feedback regulation by astromicin itself. Alternatively, inhibitory substances accumulated during the culture was probably absorbed together with astromicin.

Viscosity of the culture broth dramatically increased with the growth of producing strains like other cultures of fungi or actinomycetes. High viscosity decreased the oxygen mass-transfer coefficient (KLa) and caused incomplete mixing in the culture. To solve these problems, sterilized water was added during the culture (Figure 6). Investigation of the culture conditions from the view-

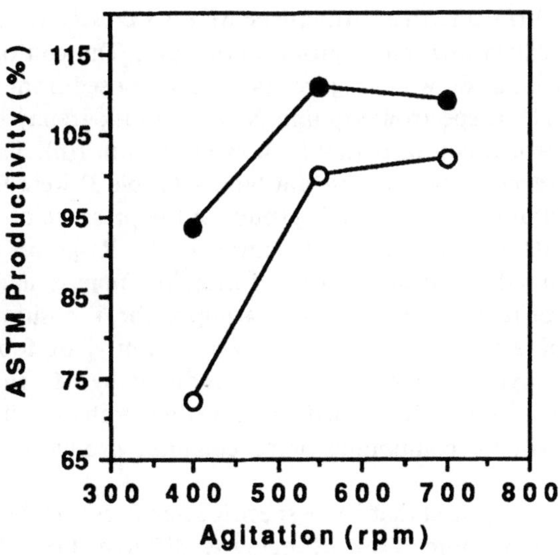

Figure 6. Effect of agitation speed on astromicin production with (●) or without (○) of addition of sterilized water during cultures.

point of dissolved oxygen levels was also an effective means to increase the production in large fermentors.

1.2. Molecular Biological Approaches for the Production of Astromicin and Its Related Antibiotics

1.2.1. Astromicin, Its Related Antibiotics, and Our Strategy

Astromicin (fortimicin A) was discovered in the screening program of products of rare actinomycetes by Nara et. al in 1977.[14] Its producer was a novel microorganism that belonged to the genus *Micromonospora*, and named as *M. olivasterospora*. Astromicin is an aminoglycoside antibiotic, having an unique structure called a pseudodissaccharide. Since then, the production of antibiotics of the same or similar structures to astromicin was reported by Japanese researchers (Table 3).[15-19] Although these producers were taxonomically different, their structures were so closely related that they were called astromicin-group antibiotics. The conformation of astromicin, as shown in Figure 7(a), was proposed from the data of proton-NMR spectrometry and X-ray crystallography of fortimicin B,[20] whose conformation is shown in Figure 7(b). The structural differences between these antibiotics (Table 3) were found in the configuration of the 1-NH_3 group, in the presence or absence of the 2-OH group and the structure of the R_4-group (glycine or N-formimidoyl glycine) on the fortamine moiety, and in the attachment site (C or N) for methyl group at the 6′ position of the purprosamine moiety [Figure 7(a)]. Very recently, we found that N-formimidoyl astromicin was also produced by *M. olivasterospora* (unpublished). It seemed to be highly probable that these formimidoyl-type compounds were common products of these producers.

It is well recognized that some specific secondary metabolites are produced by actinomycetes belonging to different taxa. Tables 4 and 5 show our data. Also, only limited strains in the taxa produce secondary metabolites. This wide but "skewed" distribution of the

Table 3. Structures of astromicin-group antibiotics [R_{1-6}; see Fig. 7(a)].

Strains		Products	R_1	R_2	R_3	R_4	R_5	R_6
Micromonospora olivasterospora	ATCC21819	Astromicin	NH_2	H	OH	$COCH_2NH_2$	CH_3	H
		Fortimicin B	NH_2	H	OH	H	CH_3	H
Micromonospora sp. SF-2098	ATCC31580	SF-2052	NH_2	H	OH	$COCH_2NHCH{=}NH$	CH_3	H
Streptomyces tenjimariensis	ATCC31603	Istamycin A	NH_2	H	H	$COCH_2NH_2$	H	CH_3
		Istamycin A_0	NH_2	H	H	H	H	CH_3
Streptomyces sannanensis	IFO14239	Sannamycin	NH_2	H	H	$COCH_2NH_2$	H	CH_3
Dactylosporangium matsuzakiense	ATCC11374	Dactimicin	NH_2	H	OH	$COCH_2NHCH{=}NH$	CH_3	H
Saccharopolyspora hirsuta	ATCC11378	Sporaricin A	H	NH_2	H	$COCH_2NH_2$	CH_3	H
		Sporaricin B	H	NH_2	H	H	CH_3	H

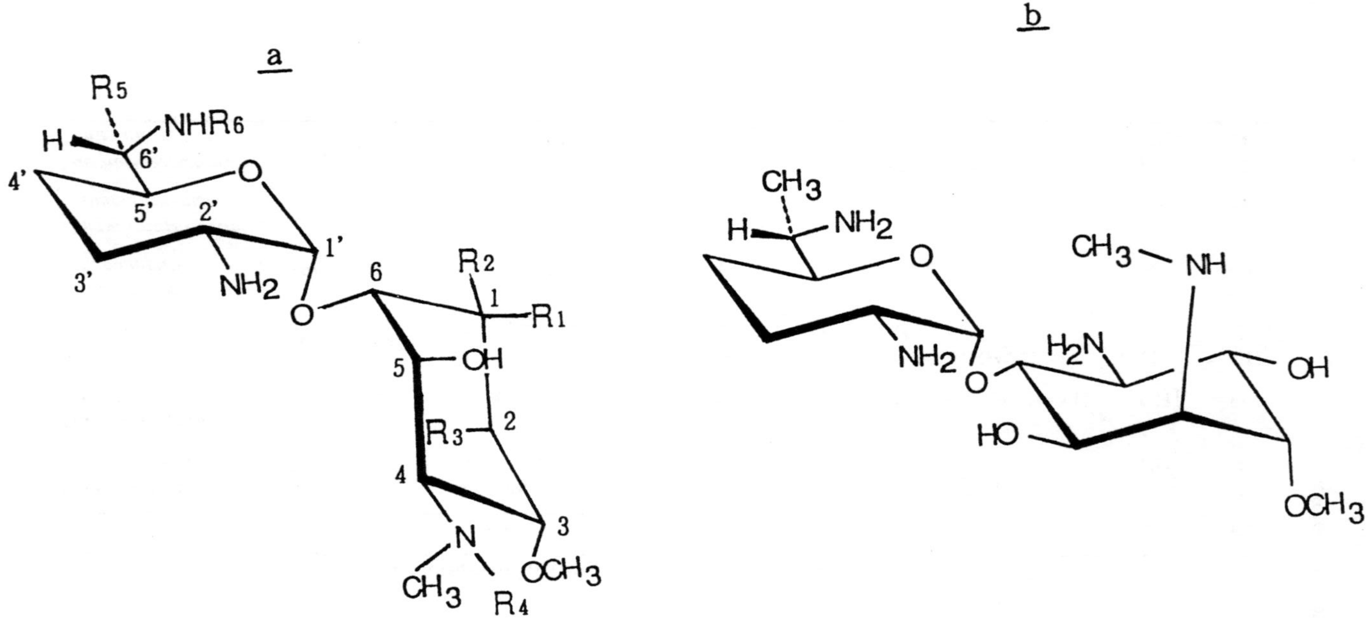

Figure 7. Structures of astromicin-group antibiotics (a) and fortimicin B (b). Structures of R_1, R_2, R_3, R_4, R_5, and R_6 are indicated in Table 3.

Table 4. Examples of antibiotics produced by specifically different Streptomyces.

Antibiotics	Producers
Actinomycins	*Streptomyces antibioticus*
	S. citreofluorescens
	S. chrysomallus subsp. chrysomallus
	S. flaveolus
	S. regensis
Mitomycins	*Streptomyces. caespitosus*
	S. cellostaticus
	S. reticuli subsup. shimofusaensis
(Hydroxy)Streptomycin	*Streptomyces bikiniensis*
	S. glaucescens
	S. griseus subsp. griseus
	S. humidus
	S. ornatus
	S. subrutilus
Tubercidin	*Streptomyces tubercidicus*
	S. cuspidosporus
	S. sparsogenes

Table 5. Antibiotics produced by generically different actinomycetes.

Antibiotics	Producers	Antibiotics	Producers
Actinomycins	*Streptomyces*	Streptomycins	*Streptomyces*
	Micromonospora		*Streptoverticillium*
Aureothricin	*Streptomyces*	Mitomycins	*Streptomyces*
	Streptoverticillium		*Streptoverticillium*
Azaserine	*Glycomyces*	Neomycins	*Streptomyces*
	Microtetraspora		*Micromonospora*
Carbomycin	*Streptomyces*	Paromomycin	*Streptomyces*
	Streptoverticillium		*Streptoverticillium*
Chloramphenicol	*Streptomyces*	Rifamycins	*Amycolatopsis*
	Streptoverticillium		*Micromonospora*
Enteromycin	*Streptomyces*	Tubercidin	*Streptomyces*
	Streptoverticillium		*Streptoverticillium*
Indolemycin	*Streptomyces*	Viomycin	*Streptomyces*
	Streptoverticillium		*Streptoverticillium*

capability to produce specific secondary metabolites peaked our interest in the genetic background of this phenomenon. Astromicin-group antibiotics are typical examples for this, because they are produced by various actinomycetes of different genera (Table 3). A taxonomic difference at the species level is sometimes confusing and ambiguous in actinomycetes, but at the generic level, it must certainly tell us something. Thus, the antibiotics of this group are good tools with which to study the distribution and evolution of secondary metabolisms in actinomycetes. As is shown later, the biosynthetic pathway of astromicin is composed of more than 14 steps. The complicated structures and the multistep biosynthesis of these antibiotics strongly suggest that they are evolutionally related, and come from a common ancestor. If this is the case, nucleotide sequences, regulation mechanisms of their genetic systems, and the enzymology of their biosynthesis must be common and alternative, admitting some minor strain-specific modifications.

Recent progress in recombinant DNA technology for actinomycetes enabled us to apply the techniques to rational strain improvement and to create hybrid antibiotics. The expected community of genetic systems and enzymatic properties for the production of antibiotics of this group greatly helps in cloning these biosynthetic genes and in applying cloned genes to manipulate these producers.

1.2.2. Biosynthetic Pathway of Astromicin

As mentioned in Sec. 1.1, many minor components of astromicin were found in the culture broths. More than 50 structures have been determined. By comparing these structures, a rudimentary idea on the pathway of astromicin biosynthesis can be determined.

The biosynthetic route was studied by bioconversion experiments of astromicin precursors and cosynthesis studies using blocked mutants of *M. olivasterospora*.[21,22]

1.2.2.1. Isolation of Blocked Mutants and Their Characters

The spores or fragmented mycelia of *M. olivasterospora* were treated with Co^{60} and EMS (methanesulfonic acid ethyl ester). Out

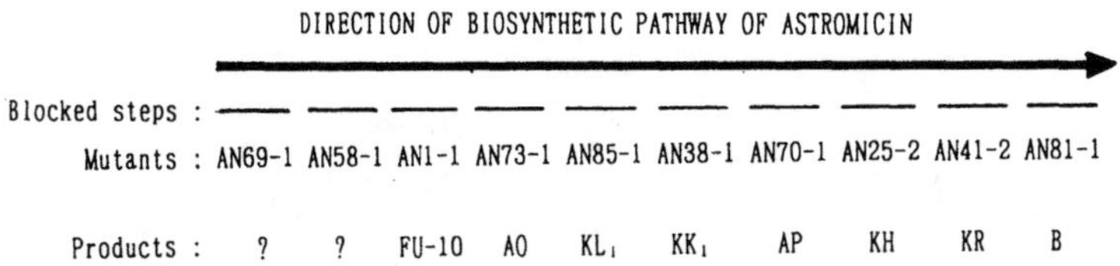

Figure 8. Characters of the blocked mutants (AN) of *M. olivasterospora*: their major products and blocked steps in the astromicin biosynthesis revealed by cosynthesis experiments.

of about 53,000 colonies 182 were found to be nonproducers of astromicin. Twenty-nine mutants were served to the cosynthesis studies[22] on agar plates, and the positions of their damaged step were surveyed. In the experiments of 812 pairs to cosynthesis, 25 mutants gave clear results, and were classified into 10 groups with respect to their blocked steps. As shown in Figure 8, representative mutants of each group were lined up in the order in which the biosynthetic pathway might be reflected. These mutants produced astromicin in liquid media when they were cocultured with mutants in the other groups. The major products (precursors of astromicin) of these mutants were isolated and identified, as shown in Figure 8. These results meant that the pathway proceeded in the order of scyllo-inosamine, FU-10, AO, KL_1, KK_1, AP, KH, KR, B, A (Figure 9). The structure of FU-10 indicated that the precursor came from the combination of scyllo-inosamine and D-glucosamine. AN69-1 acted as the converter[23] to all of the AN mutants in the 10 groups, suggesting that its defect biosynthetic step was the most earliest one in these mutants.

1.2.2.2. Bioconversion of Precursors

Washed mycelia of AN58-1 could convert precursors FU-10, AO, KK_1, AP, KH, KR, and B to an antibacterially active compound that was identified as astromicin itself by bioautogram and proton NMR. When AN81-1 was used as the catalyst, KK_1, KH, and KR were changed into B. The conversion of B to astromicin was thought to involve the glycylation of the 4-aminomethyl group of fortamine

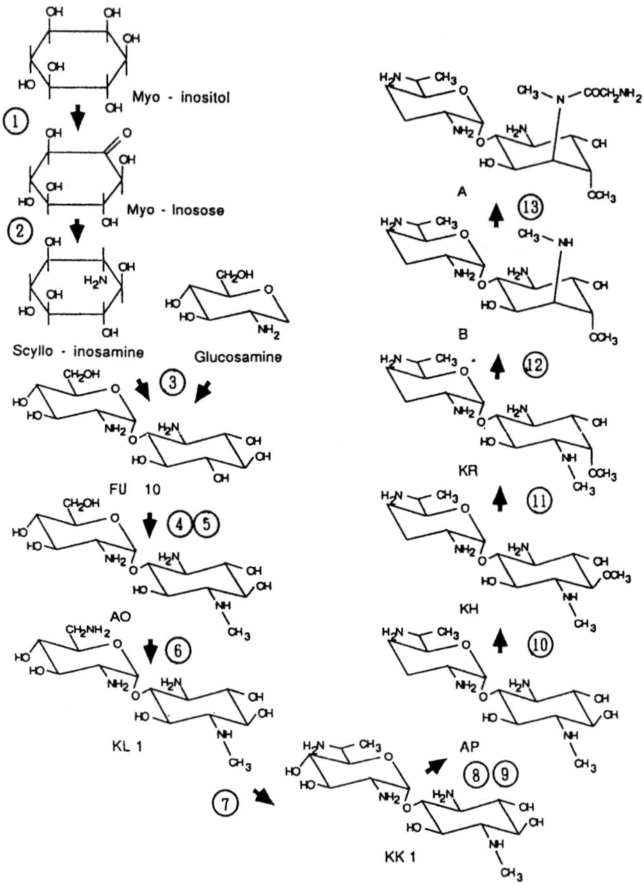

Figure 9. Proposed biosynthetic pathway of astromicin.

in B. The incorporation of glycine to astromicin was confirmed by C^{14}-glycine used as the tracer.[21]

In the proposed biosynthetic pathway, Fu-10, AO, and KK_1 were to be subjected to methylation to form AO, KL_1, and AP, respectively (Figure 9). By using L-[methyl-^{14}C] methionine, it was found that all of these methyl groups incorporated came from the group of L-methionine.

AN69–1, the blocked step that had been thought to be the most upstream of the biosynthesis was found to produce astromicin in the presences of scyllo-inosose and scyllo-inosamine, but not to produce the antibiotic from myo-inositol. Thus, a biosynthetic route earlier than FU-10 was proposed, as shown in Fig. 9. Therefore, the formation of fortamine, a novel aminocyclitol found in astromicin, was revealed to involve the common pathway for the biosynthesis of the streptidine moiety of streptomycin.[24-28]

The most distinctive nature of astromicin biosynthesis is that several important modifications of the molecule occur after the formation of the disaccharide structure. 4-Amination of the fortamine moiety occurs to form diaminocyclitol after FU-10 is formed (Figure 9). In contrast, the amination to form 1,3-diaminocyclitol occurs before the formation of frameworks for kanamycin, gentamicins, sagamicins, and butyrosin.[29-32] Furthermore, the fortamine moiety accepts at least 5 modifications in later steps (Figure 9). The only reported case for the modification of aminocyclitol after the framework formation is the acetylation of the NH_2 group in butyrosin biosynthesis.[30]

Our recent studies showed that *M. olivasterospora* could form N-formimidoyl astromicin from astromicin and glycine (results unpublished).

1.2.3. Community of the Biosynthetic Pathway Between the Producers of Astromicin-Group Antibiotics

Except for a brief report on the final step of istamycin biosynthesis,[33] the studies on the biosynthetic route of the other astromicin group antibiotics were not reported. As described earlier, we expected that these antibiotics must be synthesized through a very closely related pathway admitting some small differences depending on the producers. Regarding the use of the results of this study for gene manipulation, we chose the producers belonging to *Streptomyces*, the gene-cloning systems of which had been extensively studied.[34] We used *S. sannanensis*, the producer of sannamycin [Figure 7(a)], and tested if the strain could convert astromicin precursors (products of *M. olivasterospora*) to the intermediates downstream in astromicin biosynthesis.[35]

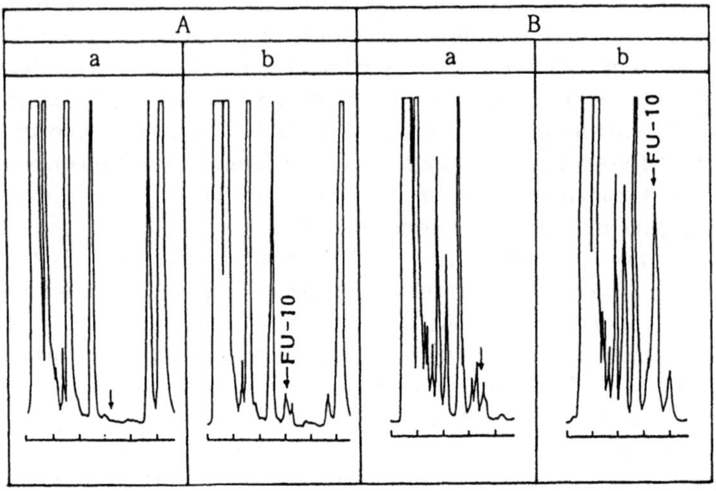

Figure 10. Bioconversion of syclio-inosamine and D-glucosamine by intact cells of *M. olivasterospora* (A) and *S. sannanensis* (B) monitored by HPLC. a: No substrates (control); b: with substrates (inosamine plus glucosamine).

The reaction mixture consisted of washed mycelium, astromicin precursors, glucose, and phosphate buffer, pH 7.1. After incubation at 30°C for 20 hours, the reaction mixtures were applied to ion pair–reverse phase HPLC. Thin-layer chromatography and proton-NMR spectrometry were performed, if necessary. Because scyllo-inosose and scyllo-inosamine were not or only poorly retained in the stationary phase of HPLC, step 1 and 2 could not be examined. Step 3 was thought to be a very important reaction where D-glucosamine was incorporated into the pathway. This is the second branching step for primary metabolism and secondary metabolism (the first branching step is step 1, where the primary metabolite myo-inositol goes into the astromicin biosynthesis). When scyllo-inosamine and D-glucosamine were incubated with *S. sannanensis*, a large amount of FU-10 was formed, as shown in Figure 10. Mycelia of 24-hour cultivation exerted the highest activity in both cases for *S. sannanensis* and *M. olivasterospora*. Those of 48 and 62 hours possessed only about one-tenth the activity. This contrasts remarkably to the other steps, which exerted rather more potent activity in the later cultivation phase. More interestingly, *S. sannanensis* showed more than fifteen times efficient converting activity than *M. olivasterospora*.

When FU-10 was used as the substrate, *S. sannanensis* converted it into AO. Thus, the strain was revealed to have both step 4 (4-amination) and step 5 (4-N-methylation), the enzymes for which could react with astromicin precursors. Formation by *S. sannanensis* of AP and KH was detected from substrate AO (Fig. 11), suggesting that the enzymes for 6′-amination (step 6), 6′-C-methylation (step 7), double dehydroxylations in the purproamine moiety (steps 8 and 9), and 3-O-methylation (step 10) were present

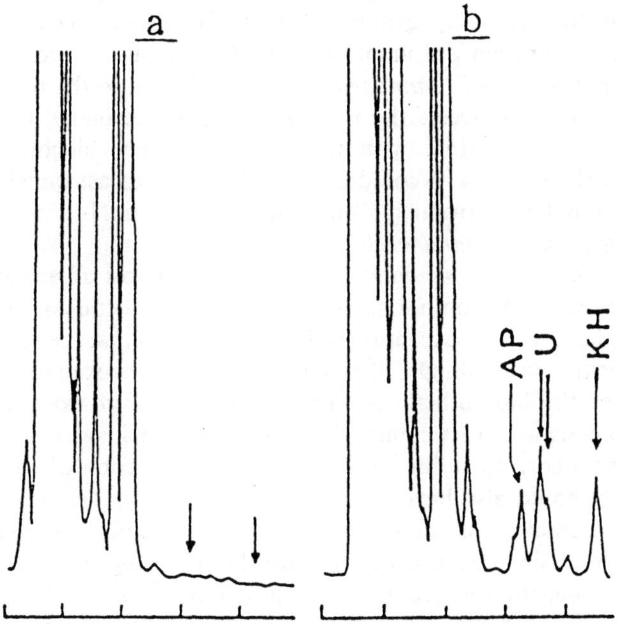

Figure 11. Bioconversion of fortimicin-AO by *S. sannanensis* monitored by HPLC. a: No substrate (control); b: with substrate (AO); U: unknown peaks specific to *S. sannanensis*.

in the stain. The precursors of astromicin could be the good substrates of these enzymes. In this case, large unknown peaks appeared in HPLC (Figure 11). We supposed that the compounds were the intermediates specific to sannamycin biosynthesis, whose structure around the 6′-position [Figure7(a)] was *Streptomyces* type (R_5=H, R_6=CH_3). Although *M. olivasterospora* could convert KH into B and A (steps 11 to 13), *S. sannanensis* was found to lack this activity. When KR was subjected to the reaction, the mycelia of *S. sannanensis* formed B and A, clearly indicating the strain had enzymes for the epimerization of 4-N-methyl group (step 12) and glycylation (step 13) of both sannamycin and astromicin precursors. These results strongly suggested that the enzyme of step 11 (epimerization of the 3-O-methyl group of the fortamine in case for astromicin) of *S. sannanensis* could not act on the astromicin precursor KH. KH possesses *hydroxyl group* at the neighboring position of the 3-O-methyl group instead of *hydrogen*, which appears in the sannamycin precursor [Figure 7(a)]. Thus, the correspondent epimerase of *S. sannanensis*, which adapted to the hydrogen atom, could not access the substrate KH, whose bulky hydroxyl group might interfere with this access. Further bioconversion studies showed that B could be transformed into astromicin and N-formimidoyl astromicin, and astromicin into N-formimidoyl astromicin by *S. sannanensis*.

In conclusion, *S. sannanensis* can convert almost all astromicin precursors to the compounds in the later biosynthetic pathway of astromicin, with the exception of KH. This means that the substrate specificities (accordingly, the structures of the pockets of active sites) of the biosynthetic enzymes of *M. olivasterospora* and *S. sannanensis* were very similar. The genetic systems, and even DNA sequences corresponding to the antibiotic production of the both strains, should also have an extended community. This may be a general condition for all producers of astromicin-group antibiotics. This conclusion encouraged us to do the following studies: transfering some specific genes among these producers to enhance the production or to create hybrid antibiotics, and cloning biosynthetic genes in the producer of interest (e.g. *M. olivasterospora*) using DNA homology with probe genes previously cloned in the other producers (e.g. *S. sannanensis*; see what follows).

1.2.4. Construction of a Gene-Cloning System of *S. sannanensis*

In actinomycetes, development and application of the recombinant DNA technique is limited to genus *Streptomyces*.[34] Besides our attempt to construct a system for *M. olivasterospora*, an astromicin producer, we succeeded in the development of a host-vector system of *S. sannanensis*, a sannamycin producer. (The idea for the study of this line came from the conclusion described in Section 1.2.3.)

1.2.4.1. Host Candidate 1, *S. tenjimariensis*

S. tenjimariensis produces istamycin, whose structure is the same as that of *S. sannanensis* [Figure 7(a)]. In the *Streptomyces* system, plasmid vectors are introduced into protoplasts by a polyethyleneglycol (PEG)-assisted manner.[34] The efficiency of regeneration of protoplasts is also critical. The protoplasting/regeneration of *S. tenjimariensis* was easily accomplished according to the method of *S. lividans*.[34] The regeneration ratio was more than 20%, sufficient for shotgun cloning. Unfortunately, this microorganism harbored more than six intrinsic plasmids. We tried to cure them by protoplasting/regeneration. This procedure was very effective and the repeated protoplasting/regeneration cured five of them. The cured strain, which harbored one last plasmid, still kept the ability to produce istamycin. All our attempts to remove the last plasmid ended in failure. The resultant strain could not be transformed by any of the plasmid vectors tested.

1.2.4.2. Host Candidate 2, *S. sannanensis*

S. sannanensis did not possess any detectable plasmids, and was sensitive to various antibiotics other than aminoglycosides. We concentrated our efforts to this microorganism to make it our cloning host. The general protocol of protoplasting/regeneration and widely used plasmid vectors[34] were not applicable to the strain. Thus, many factors for the transformation of the strain had to be examined.

For protoplasting, the strain was precultured in YEME medium[34] at 30°C for 2 days, and then cultivated further in SP4 medium (1.2% yeast extract, 2.0% peptone, 1.0% malt extract, 1.0% soluble starch, 0.5% glucose, and 0.2% glycine, pH 7.2) for 2 to 3 days. The high concentration of sucrose used for *S. lividans*[34] was inhibitory to the strain. Washed mycelia were suspended in PS medium (10.3% sucrose, 0.025% K_2SO_4, 0.005% KH_2PO_4, 2 ml/ml trace elements, 10 mM $MgSO_4$ M6H_2O, 5 mM TES, and 2.5 mH $CaCl_2$) and then were treated with lysozyme to form protoplasts. As Cu^{2+} and TES (good buffer) were found to be toxic to the protoplasts, they were removed or kept at very low concentrations in the PS medium. Very limited conditions allowed these protoplasts to regenerate efficiently. The established regeneration medium A101 consisted of 0.15% yeast extract, 0.15% malt extract, 0.01% Casamino acid, 0.1% soluble starch, 0.5% glucose, 10.3% sucrose, 1 ml/l trace elements, 0.005% KH_2PO_4, 0.025% K_2SO_4, 70 mM $CaCl_2$, 25 mM $MgCl_2$, 25 mM TES, and 2.0% Bacto agar. The pH set at 6.5 was critical. Regeneration at other pHs gave us disparate results. TES-resistant variants gave good regeneration ratio of about 10%.

Generally used vectors pIJ702 and pIJ41[34] did not go into the strain. The vector of our making, pEN101 (Fig. 12), was found to be kept stably in *S. sannanensis*. PEG4000-assisted transformation by pEN101 was highly efficient ($2.7 \times 10^6/\mu g$ DNA), which enabled us to apply this system to the shotgun cloning of sannamycin biosynthetic genes.

1.2.5. Cloning of Genes for Sannamycin Biosynthesis

1.2.5.1. Isolation of Blocked Mutants

Spore of *S. sannanensis* were treated with UV and nitrosoguanidine (NTG). Blocked mutants were obtained more efficiently when NTG was used as the mutagen. Using the ability of *S. sannanensis* to catalyze the conversion of astromicin precursors as described before, we could elucidate the blocked steps in these nonproducing mutants. We could get various mutants whose defects were at each step in the biosynthesis. Among them, mutant SN13, which failed to convert B to astromicin, was first chosen as the cloning host for the gene (*sms*13) coding for glycyltransferase (step 13).

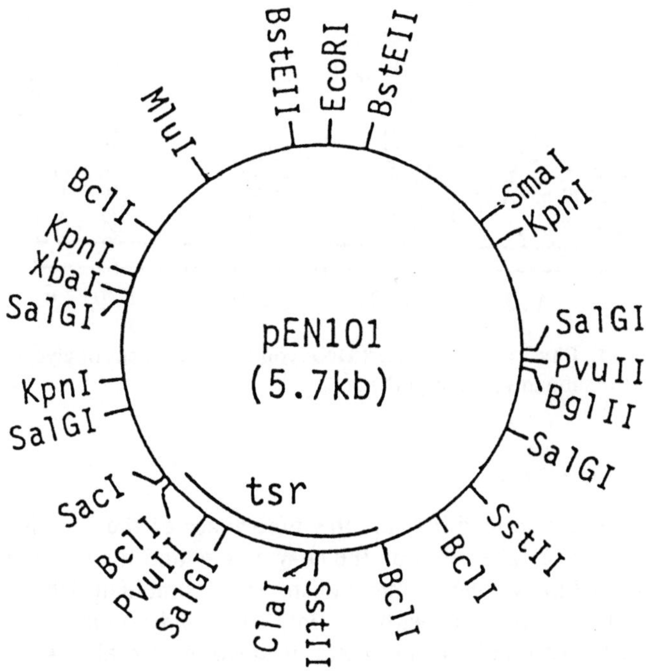

Figure 12. Structure of vector pEN101 for *S. sannanensis.*

1.2.5.2. Shotgun Cloning of sms 13

The total DNA of *S. sannanensis* was partially digested with *Sau*3AI, and 4–6-kb fragments were collected. These fragments were integrated into the *Bgl*II site of pEN101, and the resultant chimeric plasmids were introduced into the protoplasts of SN13. Out of about 6000 regenerated colonies, one recovered the sannamycin productivity. Its plasmid was isolated, and the physical map of the plasmid was determined, as shown in Figure 13.[36]

Sms 13 gene in this insert DNA has been successfully used as the probe to clone the corresponding *Micromonospora* gene fms 13 from its cosmid library (in preparation for publication).

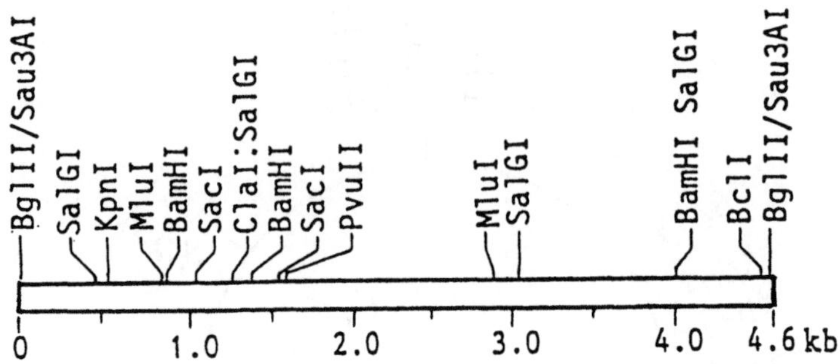

Figure 13. Physical map of insert DNA containing the gene of glycyltransferase for sannamycin biosynthesis.

1.2.6. Conclusion

The studies described indicate the importance of the combination of rather classical studies on the enzymology and the biosynthetic pathway of the secondary metabolism and recombinant DNA technique. They provide us with not only basic understandings of the genetic backgrounds of antibiotic production, but also with new tools and novel ideas for industrial production.

CHAPTER 2

Development of Sedecamycin and Application of Cyclodextrin in the Fermentative Production Process

Takashi Suzuki and Yoshiharu Tsubota

2.1. Introduction

Since penicillin was discovered by Fleming in 1929, over 4000 antibiotics have been isolated, and more than 30 000 compounds have been derived from those natural antibiotics.

Among these, about 100 compounds are used clinically in the field of infectious diseases, but, recently, antibiotics have also been used in the fields of animal and plant diseases in addition to the medical field.

In the field of animal diseases, many antibiotics are used in Japan. Tetracycline-group antibiotics are widely used for growth promotion or prevention of infectious diseases. Macrolide antibiotics such as tylosin are used for increased weight gain in swine or improved feed efficiency in chickens. Polyether antibiotics such as salinomycin and monensin aré used as anticoccidiostats for poultry and feed additives for cattle. Colistin and bacitracin are peptide antibiotics that are produced by *Bacilli* and are used as animal-feed additives in Japan. Virginiamycin, enramycin, thiopeptin, and bicyclomycin, belonging to the same category, are produced by *Streptomyces* and are also used as animal-feed additives. Destomycin and hygromycin are basic oligosaccharide antibiotics that are active againt animal parasites and used as animal-feed additives.

Among those antibiotics used in the field of animal disease, salinomycin, colistin, enramycin, thiopeptin, bicyclomycin, and destomycin were discovered and developed in Japan.

Although many antibiotics have been developed and are widely used in animals, it is very important to check carefully for problems related to residual toxicity, environmental pollution, and cross-resistance of the antibiotics in addition to checking their effectiveness.

We recently developed a new macrolide antibiotic, sedecamycin, which is obtained by the fermentation of *Streptomyces rochei* var. *volubilis*. It is an effective and safe drug for swine dysentery, which has become a serious problem in the livestock industry.

First, it was thought that producing sedecamycin on an industrial scale would be difficult, because the amount of accumulation in the media was very low, owing to the characteristics of the antibiotic such as hydrophobicity and instability in water. However, the reseachers discovered that the addition of cyclodextrin resulted in the formation of an inclusion complex that is hydrophilic and stable in water. The industrial production of sedecamycin was realized on the basis of these results.

This chapter describes the development of sedecamycin and the application of cyclodextrin to the fermentative production process.

2.1.1. List of Abbreviations

The following common abbreviations are used in this chapter.

CyD:	Cyclodextrin
^{14}C-Met:	L-(^{14}C-methyl)-methionine
DS-medium:	The medium containing dextrin and defatted soybean flour.
GP-medium:	The medium containing glycerol and polypepton.
HON:	DL-δ-hydroxy-γ-oxo-norvaline
HPLC:	High-performance liquid chromatography
TLC:	Thin-layer chromatography
PEG:	Polyethylene glycol
MIC:	Minimum inhibitory concentration

2.2. Chemical Structure of Sedecamycin

2.2.1. Sedecamycin

Sedecamycin was obtained by the fermentation of a *Streptomyces* strain that was isolated from a soil sample collected in Nukata,

Osaka, Japan, followed by enzymatic acetylation[37] and extraction and purification.[38] The strain was named as *Streptomyces rochei var. volubilis* by Higashide et al.[37]

The antibiotic is active against gram-positive bacteria, has low toxicity, has no cross-resistance with known antibiotics, and is effective even when administered orally (shown in Sec. 2.3). Component substances were isolated from the cultures of the previous strain and were initially named T-2636 A, C, and so on, after the type culture deposit serial number (T-2636) of the producing strain.[39-41]

A subsequent study revealed that this group of antibiotics has a specific 17-membered ring composed of a lactone ring and a carbon–carbon bond structure, including a β-keto-δ-lactone ring, and that T-2636 A is identical to lankacidin A and T-2636 C to lankacidin C.[42] Sedecamycin was selected as the Recommended International Non-Proprietary Name in 1986, because the antibiotic has a 17-membered ring and the Latin word for 17 is septem-decin. Hereafter, we call the antibiotic sedecamycin in place of T-2636 or lankacidin.

Swine dysentery, one of the most important diseases in swine, is characterized by a severe mucohemorrhagic diarrhea and is rampant in virtually all parts of the world. Takeda conducted exploratory research to develop a new, effective, and safe drug for this disease. In the course of research, it was discovered that sedecamycin has remarkable inhibitory effects on the causative agent, *Treponema hyodysenteriae*, not only *in vitro*, but also *in vivo* and these are described in Sec. 2.3.

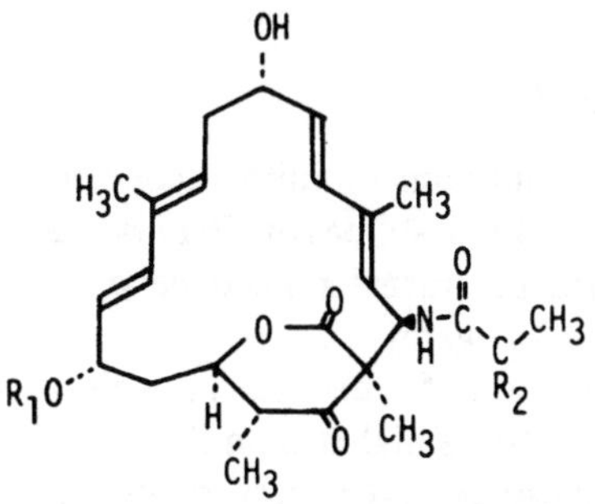

	R_1	R_2
Sedecamycin C	-H	=O
Sedecamycin F	-H	-H, -OH
Sedecamycin A	-Ac	=O
Sedecamycin D	-Ac	-H, -OH

Figure 14. Structure of sedecamycins.

Sedecamycin, which is isolated using the methods described in Sec. 2.5., consists of mainly sedecamycin A (more than 95%) and only a small portion is made up of other components (sedecamycin, C, D, E). The chemical structure of sedecamycin A, which is shown in Figure 14, can also be expressed as (-)-N-[1S, 2R, 3E, 5E, 7S, 9E, 11E, 13S, 15R, 19R)-7, 13-Dihydroxy-1,4,10,19-tetramethyl-17, 18-dioxo-16-oxabicyclo [13.2.2] nonadeca-3, 5, 9, 11-tetraen-2-yl] pyruvamide 13-acetate. Sedecamycin A is the same as lankacidin A or T-2636A.

2.2.2. Related Compounds of Sedecamycin

Streptomyces rochei var. volubilis produces a small amount of sedecamycin F in addition to sedecamycin C, when fermentation is carried out at a low level of dissolved oxygen.

Sedecamycin F can be converted to sedecamycin D and sedecamycin C to sedecamycin A, if the products in the broth are acetylated enzymatically in the presence of an acetyl donor such as ethyl acetate. Carboxyl esterase, which is produced by sedecamycin-producing strains, catalyzes not only acetylation, but also deacetylation of these sedecamycins.[43]

Harada et al.[39] and Kamiya et al.[42] reported the isolation and structures of these four sedecamycins. Fugono et al. described the enzymatic interconversions of these antibiotics.[43]

2.3. Development of Sedecamycin

2.3.1. Antimicrobial Activity

2.3.1.1. In Vitro Antimicrobial Activity

Yamazaki et al.[44] reported that the minimum inhibitory concentration (MIC) of sedecamycin against 79 field isolates of *Treponema hyodysenteriae* obtained from several pig farms ranged between 0.78 and 12.5 μg/ml with a peak of 3.13 μg/ml.

The minimum bactericidal concentration of sedecamycin against *T. hyodysenteriae* was observed to be 4 times higher than the MIC. Morphological changes in *T. hyodysenteriae* were observed under a

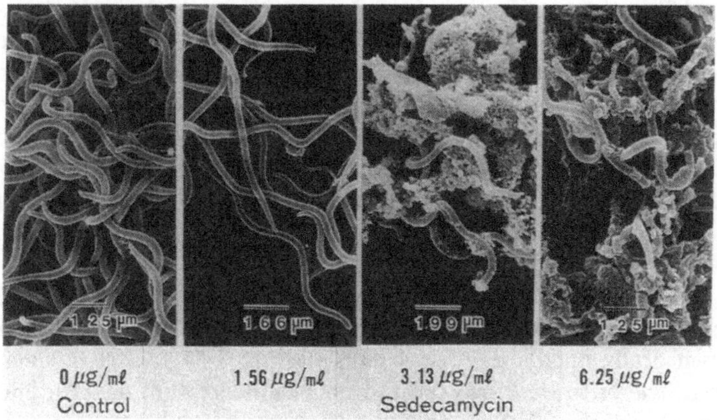

Figure 15. Scanning electron micrographs of *T. hyodysenteriae* exposed to sedecamycin.

scanning electron microscope (Figure 15) when the cells were incubated in liquid media containing sedecamycin. *T. hyodysenteriae* retained its shape up to 1.56 μg/ml, but morphological changes such as transformation and lysis of the cell and the outflow of the cellular content were observed above 3.13 μg/ml. It was therefore morphologically demonstrated that sedecamycin affects *T. hyodysenteriae* bactericidally at concentration above the MIC.

The MICs of sedecamycin against various bacteria were first reported by Tsuchiya et al.[45] Sedecamycin shows strong antibacterial activity against Gram-positive bacteria, but does not have activity against Gram-negative enterobacteria. Sedecamycin, however, shows strong activity against *Pasteurella* and *Haemophilus*, which are involved in respiratory diseases in cattle and swine. Sedecamycin inhibits anaerobic *Fusobacterium necrophorum*, which is known to influence the severity of swine dysentery, at the concentration of 12.5 μg/ml.

2.3.1.2. Experimental Infection Study in Mice

The protective effects of sedecamycin against experimental infection caused by *T. hyodysenteriae* in mice were reported by Hayashi et al.[46] in comparison with reference antimicrobials. Test drugs were given orally on day 1 and day 2 after inocculation. The mice were sacrificed on day 7 after inocculation, and gross cecal lesions and the colonization of *T. hyodysenteriae* in the cecum were examined. Sedecamycin showed a strong protective effect: the ED_{50} was 7.0 mg/kg. The ED_{50}s of carbadox, tiamulin, and lincomycin were 7.7, 14.6, and 21.7 mg/kg, respectively. Sedecamycin was an active as carbadox and two or three times more effective than tiamulin and lincomycin.

Mice were challenged with *T. hyodysenteriae* to provide a model for studying the therapeutic effect of sedecamycin. Test drugs were given orally once a day for 4 days starting 7 days after inoculation. The mice were necropsized on postinoculation day 14, and the colonization of *T. hyodysenteriae* in the cecum was examined. Sedecamycin was highly effective: the ED_{50} was 6.4 mg/kg. The ED_{50}s of carbadox, tiamulin, and lincomycin were 5.0, 12.1, and 12.8 mg/kg, respectively. Sedecamycin was an active as carbadox and two times more effective than tiamulin and lincomycin.

2.3.2. Effects Against Swine Dysentery

2.3.2.1. Prophylactic and Therapeutic Effects in Experimental Infection

Yamazaki et al.[8] reported the prophylactic and therapeutic effects against experimentally induced swine dysentery.

The prophylactic effects of sedecamycin were studied by feeding piglets diets containing sedecamycin 7 days prior to infection through 21 days after infection. Sedecamycin demonstrated a prophylactic effect against swine dysentery in a dose-related manner (Table 6). In the lower-dose groups, the number of pigs showing clinical signs of swine dysentery and the duration of mucohemorrhagic diarrhea were remarkably reduced, but the number of pigs shedding *T. hyodysenteriae* and the number of pigs positive for colonic lesions were not significantly different from the numbers obtained from the infected nonmedicated controls. In the intermediate-dose groups, no pigs showed any clinical signs, but a few pigs shed the organism in the feces and were positive for colonic

Table 6. Prophylactic effect of sedecamycin against experimentally induced swine dysentery.

Treatment	No. of pigs	No. with mucohemorrhagic diarrhea	Mean duration during the 21 days after infection[a]		No. positive for *T. hyodysenteriae*		No. positive for colonic lesion	Performance during the 21 days after infection[a]	
			Mucohemorrhagic diarrhea	Diarrhea	Feces	Colonic mucosa		Weight gain (g)	Feed-conversion ratio
Infected non-medicated control	12	12^{b}	12.7^{b}	16.2^{b}	12^{b}	12^{b}	12^{b}	199^{b} (37)	2.77^{b}
Sedecamycin, 10–12.5 ppm	12	4^{c}	2.3^{c}	6.5^{bc}	9^{b}	11^{b}	11^{b}	464^{c} (85)	1.91^{c}
Sedecamycin, 20–25 ppm	12	0^{d}	0^{c}	1.9^{c}	3^{c}	2^{c}	2^{c}	598^{d} (110)	1.91^{c}
Sedecamycin, 40–50 ppm	12	0^{d}	0^{c}	0.4^{c}	0^{c}	0^{c}	0^{c}	599^{d} (110)	1.77^{c}
Carbadox, 50 ppm	12	0^{d}	0^{c}	0.6^{c}	0^{c}	0^{c}	0^{c}	543^{cd} (100)	1.82^{c}

[a] Figures in parentheses indicate % of the value for carbadox 50 ppm.
[b–d] Values within columns that do not have a common superscript differ, $p < 0.05$.

lesions. In the higher-dose groups, all evaluation parameters were negative, as was the case in the carbadox group. All medicated groups showed significantly greater weight gain, and the feed conversion ratio was also significantly improved. The sedecamycin intermediate-and higher-dose groups showed better growth performance than did the carbadox group.

The therapeutic effects of sedecamycin were studied by challenging piglets with *T. hyodysenteriae* to induce swine dysentery experimentally and continuously thereafter, feeding a diet containing sedecamycin. Sedecamycin at 6.25 ppm showed hardly any therapeutic effect, whereas dose-dependent efficacy was observed from 12.5 to 75 ppm (Table 7). The efficacy at 100 ppm was similar to that at 75 ppm. No remarkable differences in drug efficacy were noted between the laboratory strain and the field isolates. Dose-related efficacy was observed for each evaluation parameter. It should be noted that Table 7 shows the number of days to recover to normal feces, and, therefore, mucohemorrhagic diarrhea had already disappeared by then.

2.3.2.2. Field Trials

Field trials with sedecamycin were conducted by treating pens that contained infected pigs. An example of the results of such a trial was reported by Ishikawa et al.[47] The trial was conducted on a pig farm that was raising about 1000 fattening pig and was contaminated with swine dysentery. Sedecamycin was added to the feed for the infected pens at several concentrations for 10 days. A carbadox-treated group and an untreated group were also included. In the sedecamycin-treated groups, dose-related improvement in fecal appearance and decreased recovery of the organism were observed (Table 8). When compared with the results of experimental infection studies, the results from this trial tended to fluctuate, probably due to the presence of anorectic animals or reinfection from the adjacent pens.

2.3.3. Toxicity

The LD_{50} of sedecamycin is $> 10\,g/kg$ p.o. and s.c. in mice and rats. Sedecamycin was administered to rats for 4 weeks at levels of

Table 7. Therapeutic effect of sedecamycin against experimentally induced swine dysentery.

Conc. of sedecamycin in feed (ppm)	No. of pigs[a]	Recovery to normal feces		Disappearance of *T. hyodysenteriae*		No. of pigs[b] necropsized	Colonic mucosa at necropsy	
		No. (%)	Days	No. (%)	Days		No. positivee for *T. hyodysenteriae* (%)	No. positive for lesion[c]
12.5	9	9 (100)	4.6	8 (89)	6.4	9	4 (44)	7 (78)
25	32	31 (97)	2.6	30 (94)	4.4	27	4 (15)	3 (11)
50	38	38 (100)	2.3	38 (100)	3.1	30	0 (0)	0 (0)
75	5	5 (100)	1.4	5 (100)	2.2	5	0 (0)	0 (0)

[a] 6–9 weeks old at the time of infection.
[b] 7–10 days after the start of medication.
[c] Those that had only synulotic ulcers were excluded (2 pigs in the 25-ppm group and 8 pigs in the 50-ppm group).

Table 8. The correlation of fecal appearance and recovery of T. hyodysenteriae from feces.

Treatment (no. pigs)	Fecal appearance	Day 0 −	Day 0 +	Day 0 ++	Day 0 +++	Day 0 ++++	Day 4 −	Day 4 +	Day 4 ++	Day 4 +++	Day 4 ++++	Day 7 −	Day 7 +	Day 7 ++	Day 7 +++	Day 7 ++++	Day 10 −	Day 10 +	Day 10 ++	Day 10 +++	Day 10 ++++
Sedecamycin 75 ppm (19)	●	1	1			1															
	◉			1		2															
	◎	1	1	1	2						1						3				
	○	6	1	1			18					18	1				15	1			
Sedecamycin 50 ppm (22)	●		2	1		3															
	◉	1																			
	◎																1				
	○	14	1				22					22					21				
Sedecamycin 25 ppm (21)	●				1	2															
	◉	1			1	1															
	◎		1	1	1												2				
	○	7	3	1	1		20		1			21					19				
Sedecamycin 12.5 ppm (17)	●	2				4															
	◉	1		2	3																
	◎				1				1	1	1					1			1		
	○	3		1			11	1			2	15		1			13		2		1
Carbadox 50 ppm (21)	●	1			1	2				1	1					2					1
	◉	1				1				1		1								1	1
	◎						1					1				1	1	1			
	○	13	1	1			14	3				16					15	1			
Untreated control (21)	●		1	1	3	4			1	2	5			1	2	4			1		3
	◉				1	2				1							2		2		1
	◎													1	1	1					
	○	8			1		8	2			2	11					9	1	1		1

Fecal appearance

● : Mucohemorrhagic diarrhea

◉ : Diarrhea

◎ : Soft

○ : Normal

Number of hemolytic zones per plate

− : 0

+ : 1~9

++ : 10~99

+++ : over 100

++++ : hemolytic zone all over plate

0, 100, 300, 1000, and 3000 ppm in their diet. No death occurred in any dosage group during the experiment. Slight changes in blood chemistry were observed in the 1000 ppm group and body weight gain was reduced in the 3000 ppm group.

Sedecamycin was administered to rats for 13 weeks at levels of 0, 80, 250, 700, and 2000 ppm in their diet. No death occurred in any dose group during the experiment. Although slight changes in blood chemistry were observed in the animals treated with less than 700 ppm, their weight gain exceeded that of the untreated controls. Body weight gain was reduced in the 2000 ppm group.

Teratogenicity, mutagenicity, and skin and ocular mucosal irritation were studied and were all negative.

Sedecamycin was added to the swine diet at the level of 500 ppm and was given to 2-month-old pigs continuously for 14 days. No adverse effects on clinical findings or growth were observed. No drug-related changes were observed in hematology or urinalysis or at necropsy. In another study in which a diet containing sedecamycin at 250 ppm was given for 4 weeks, no adverse effects on growth, etc., were observed.

2.3.4. Pharmacokinetics and Tissue Residues

When administered orally to pigs, sedecamycin is rapidly absorbed from the intestinal tract and distributed to the intestinal mucosa via blood circulation. A portion of the sedecamycin (30%) thus absorbed is excreted in the urine via the kidneys, and most of it is released into the intestinal tract and excreted in the feces together with the unabsorbed portion (70% excreted in feces).

Okada et al.[48,49] reported that when sedecamycin was added to a swine diet at the level of 50 ppm and given to pigs for 14 days, neither sedecamycin nor its active metabolites were detected in edible tissues, even after no withdrawal period. Similarly, when sedecamycin was added to the swine diet at the level of 500 ppm and given to pigs for 14 days, trace amounts of sedecamycin or its major metabolites were detected in the liver, small intestine, and colon when there was no withdrawal period, but no residue was detected 1 day after drug discontinuation. Based on these results, it was

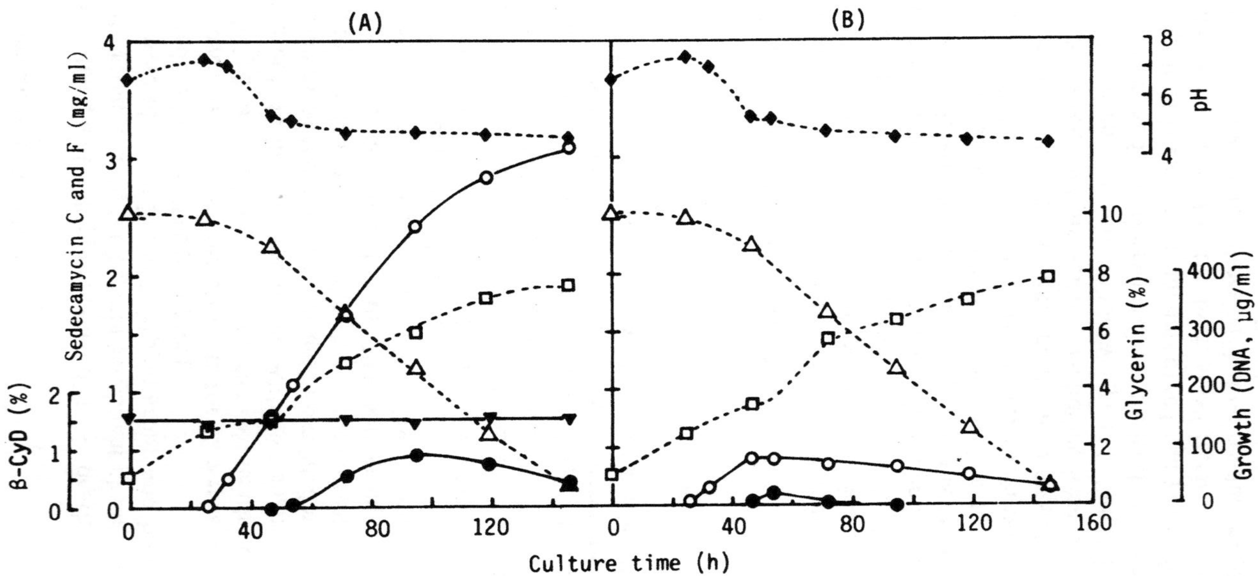
(A)
(B)
Sedecamycin C and F (mg/ml)
β-CyD (%)
pH
Glycerin (%)
Growth (DNA, μg/ml)
Culture time (h)

Figure 16. (see facing page) Time course of sedecamycin production by *Streptomyces rochei* var. *volubilis* HON^R-190 in the presence (A) and absence (B) of β-CyD (1.5%). Sedecamycins were extracted from the fermentation broth with methyl isobutyl ketone and assayed by HPLC under the following conditions: column: 3.9 mm ID × 300 mm, packed with μ Poracil, 10 μ (Waters Assoc. Inc., Tokyo); mobile phase, n-hexane-isopropanol (65:35, v/v) at a flow rate of 1.2 ml/min.; detection, UV at 254 nm; temperature, 45°C. The retention times of sedecamycin C, sedecamycin F, sedecamycin A, sedecamycin D were 4.6 min., 4.8 min., 3.5 min., and 3.9 min., respectively. β-CyD was determined as follows. The twofold diluted fermentation broth was mixed with an equal volume of methyl isobutyl ketone to dissociate the inclusion complex, and the aqueous layer was used for the determination of β-CyD by HPLC under the following conditions: column, 4.0 mm ID × 250 mm, packed with Shimadzu PNH_2-10/S 2504; mobile phase, acetonitrile-water (70:30, v/v) at a flow rate of 1.2 ml/min.; detection, differential refractometer; temperature, 28°C. Microbial growth was determined by measuring the DNA in the cells. Glycerol was assayed on the basis of spectrophotometric determination of hydrogen peroxide generated by sequential enzymatic reactions (glycerol kinase, glycerol-3-phosphate oxidase) using Triglyceride G-Test Wako (Wako Pure Chem. Ind., Ltd.). □: Microbial growth; △: glycerin; ○: sedecamycin C; ●: sedecamycin F; ▼: β-CyD; ◆: pH. (Sawada et al.[51])

determined that the withdrawal period for sedecamycin should be 1 day.

2.4. Cyclodextrin Effect

2.4.1. Effect of Cyclodextrin on the Fermentative Production of Sedecamycin

Because β-Cyclodextrin (β-CyD) forms an inclusion complex with sedecamycins (Harada et al.[50]), it was expected that β-CyD might have some effects on sedecamycin production. Figure 16 shows typical time course of sedecamycin fermentation in the presence (A) and absence (B) of β-CyD.

Culture media and conditions are as follows. A spore suspension of *Streptomyces rochei var. volubilis* HON^R-190 derived from the original strain (IFO 12507) was transplanted into the seed medium. The composition of the seed medium was 2.5% glucose, 1% proflo (Trader Oil Mill Co., USA), 3.5% corn steep liquor, 0.1% K_2HPO_4, 0.02% $MgSO_4 \cdot 7H_2O$, and 1.5% $CaCO_3$, pH 6.5. The seed culture was performed at 28°C for 17 h on a rotary shaker. One milliliter of the seed culture was transferred to 25 ml of a fermentation medium in a 200-ml Erlenmyer flask. The composition of the fermentation medium was 10% glycerol, 2% Proflo, 1% polypeptone, 0.5% corn steep liquor, 0.5% NaCl, 0.1% $FeSO_4 \cdot 7H_20$,0.0025% $CuSO_4 \cdot 5H_2O$, and 0.01% Actcol (an antifoaming agent, Takeda Chemicals, Industries, Ltd.), pH 5.8, β-CyD was added to the medium before autoclaving unless otherwise stated. The fermentations was ordinarily carried out at 24°C for 6 days on a rotary shaker.

In the absence of β-CyD, only 0.4 mg/ml of sedecamycin C had accumulated after culturing for 48 h. The amount then decreased slowly. In contrast, in the presence of β-CyD, sedecamycin C production continued throughout the fermentation and finally reached 3.1 mg/ml. β-CyD has a great effect on the production of sedecamycin, but it has no marked effect on microbial growth, glycerol consumption, or changes in pH. The residual amount of β-CyD in the culture was measured to determine whether or not it is consumed by the microorganisms. As shown in Figure 16 (A), β-CyD added to the medium was not consumed during fermentation.

Next, to confirm the general effect of β-CyD, other *Streptomyces* species were examined. These species were *S. rochei* var. *volubilis*, *S. violaceoniger* NRRL 2834, *S. violaceoniger* ATCC 10976, *S. violaceoniger* ATCC 19933, *S. griseofuscus* IFO 12870, and *Streptomyces* sp. No. 6642G. These are known to produce sedecamycin-group antibiotics. As shown in Sawada et al.[51], β-CyD was effective in increasing the amount of sedecamycins accumulated in each strain mentioned before.

2.4.2. The Presence of an Inclusion Complex

To determine the presence of an inclusion complex of sedecamycin C and β-CyD in the fermentation broth, Figure 16(A), 10 ml of the supernatant solution from the fermentation broth was passed through a Cellulofine GCL-90-m column (3.1 cmID 52 cm, Seikagaku Kogyo Co. Ltd., Tokyo) with water at a flow rate of 2.8 ml/min. Each 5-ml fraction was collected and assayed for sedecamycin C and β-CyD by HPLC. β-CyD and sedecamycin C appeared in the same fraction, No. 29, Figure 17(A). When authentic samples

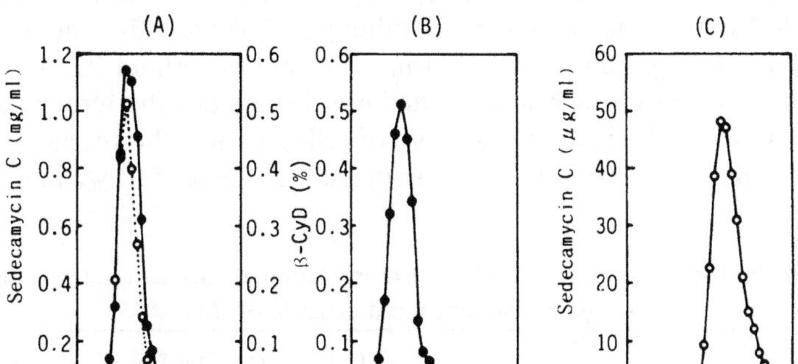

Figure 17. Occurrence of an inclusion complex of sedecamycin C with β-CyD in the culture filtrate. (A) Culture filtrate, (B) authentic sample of β-CyD, and (C) authentic sample of sedecamycin C. ○: Sedecamycin C; ●: β-CyD. (Sawada et al.[51]).

of β-CyD and sedecamycin C were separately passed through the column in the same manner, the former was eluted at a maximum level in fraction No. 39, Figure 17(B), and the latter in fraction No. 47, Figure 17(C). These results suggested that sedecamycin is accumulated in the fermentation broth as the inclusion complex with β-CyD. This idea was supported because the sedecamycin C produced was present in the culture filtrate at a much higher concentration than that expected on the basis of the solubility of sedecamycin C itself.

2.4.3. Effect on the Incorporation of L-^{14}C-Methionine into the Sedecamycin Molecule

L-(^{14}C-methyl)-methionine (^{14}C-Met) is a precursor of sedecamycin in its biosynthesis and is known to be incorporated effectively into the methyl residue of the sedecamycin molecule (Hatano et al.[52]). Furthermore, it is assumed that this incorporation is closely related to the sedecamycin-producing activity (Sawada et al.[51,53]).

To clarify the difference in sedecamycin-producing activity between the cells grown in the presence and absence of β-CyD, the incorporation of ^{14}C-Met into the sedecamycin molecule and measured in broths cultured for 1, 2, 3, and 4 days after inoculation. Table 9 shows that the culture containing β-CyD not only exhibited a higher incorporation activity than the culture without it, but it also maintained a high level of activity throughout the fermentation. On the otherhand, the activity of cells grown in the absence of β-CyD decreased rapidly. It was assumed that the cells grown in the

Table 9. Incorporation of ^{14}C-labeled methionine into sedecamycin in cultures in the presence and absence of β-CyD.

β-CyD in culture	Culture time (days)[a]			
	1	2	3	4
	Incorporation activity (× 10^4 dpm)			
1.6%	17.1	22.2	19.9	17.3
0%	10.5	6.4	1.0	0.7

[a] *Streptomyces rochei var. volubilis* HON^R-190 (described in Sec. 2.5.1) was employed.
Source: Sawada et al.[4]

presence of β-CyD became those having high activity, and it was presumed that β-CyD cancels a feedback repression of sedecamycin biosynthesis by the formation of the inclusion complex.

2.4.4. Mechanism of Cyclodextrin Effect

2.4.4.1. Effect of Various Kinds of Cyclodextrins

The effect of β-CyD on sedecamycin production by *S. rochei* var. *volubilis* was examined and compared with those of α-CyD and γ-CyD. As can be seen from Table 10, it was found that β-CyD greatly stimulated the production of sedecamcyin C. γ-CyD had an intermediate effect and α-CyD exhibited the least effect. The most effective concentration of β-CyD was 1.5% (W/V), but the most effective concentration depends on the medium.

Dimethyl and trimethyl β-CyD inhibited sedecamycin production, because they have a remarkable inhibitory effect on the growth of the sedecamycin-producing microorganism.

To clarify the difference in sedecamycin-producing activities among the cells grown in the presence of α-, β-, and γ-CyD, the incorporation of C^{14}-Met into the sedecamycin C molecule was measured in the broths cultured 2 days. As shown in Figure 18, the incorporation was greatest with cells grown in the presence of β-CyD and was less with γ-CyD. The activity of cells grown in the presence of α-CyD was the least, but was distinctly greater than that of cells without CyD. These results coincide with the data obtained

Table 10. Stimulatory effect of α-CyD, β-CyD, and γ-CyD on sedecamycin C production[a] by Streptomyces rochei *var.* volubilis *HON^R-190*[b].

CyD type	CyD concentration (%)				
	0	0.5	1.0	1.5	2.0
	Sedecamycin C production (mg/ml)				
α	0.22	0.37	0.60	0.73	0.75
β	0.22	1.69	2.61	3.00	2.83
γ	0.22	0.57	0.88	1.10	1.27

[a] After 5 days culturing at 24°C, the amount of sedecamycin C produced was estimated by HPLC.
[b] Described in Sec. 2.5.1.
Source: Sawada et al.[51]

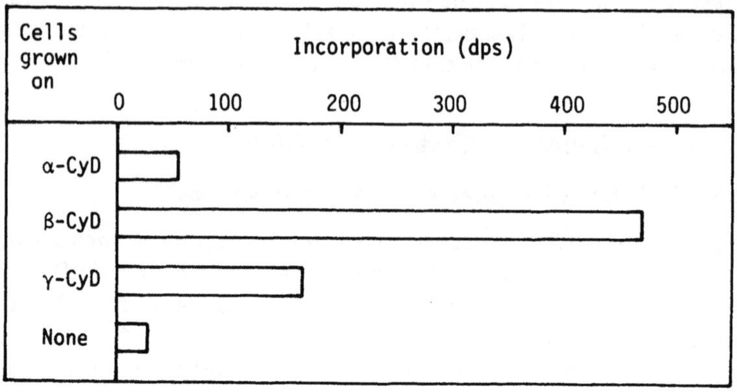

Figure 18. Incorporation of L-[methyl-^{14}C]methionine, in disintegrations per second (dps) into the sedecamycin molecule in cells grown in the presence of CyD. Culturing was carried out for 2 days in media containing α-, β- or γ-CyD or without CyD. The incorporation of L[methyl-^{14}C]methionine was measured as follows. After the cells were harvested and washed in 0.1 M 2-(N-morpholino)-ethane-sulfonic acid (MES) buffer (pH 5.6) twice, they were added to the reaction mixture at the concentration of 2 mg/ml (dry-cell weight). The reaction mixture consisted of 1% glycerol, 0.5% β-CyD, 0.1 M MES buffer (pH 5.6), and 0.5 μCi/ml L-[methyl-^{14}C]methionine. After incubation at 24°C, the filtrate was mixed with methyl isobutyl ketone (1:1, v/v), and radioactive sedecamycins were extracted in the solvent layer. Radioactivity of these fractions was measured by the method described in the previous paper.[51] (Sawada et al.[53])

on fermentative production and described before.

The affinity of α-, β-, and γ-CyD for sedecamycin C was investigated by gel filtration based on the report of Korpela and Himanen[54]. Sedecamycin C (20 mg) and each CyD (200 mg) were mixed in 20 ml of water. After shaking for 30 min. at room temperature, the mixtures were centrifuged and the supernatants obtained were passed through filters (0.45 μm). The concentration of sedecamycin C was 220 μg/ml in the filtrate without CyD, 210 μg/ml with α-CyD, 1 mg/ml with β-CyD, and 1 mg/ml with γ-CyD. These filtrates were passed through a Cellulofine GCL-90-m column (the

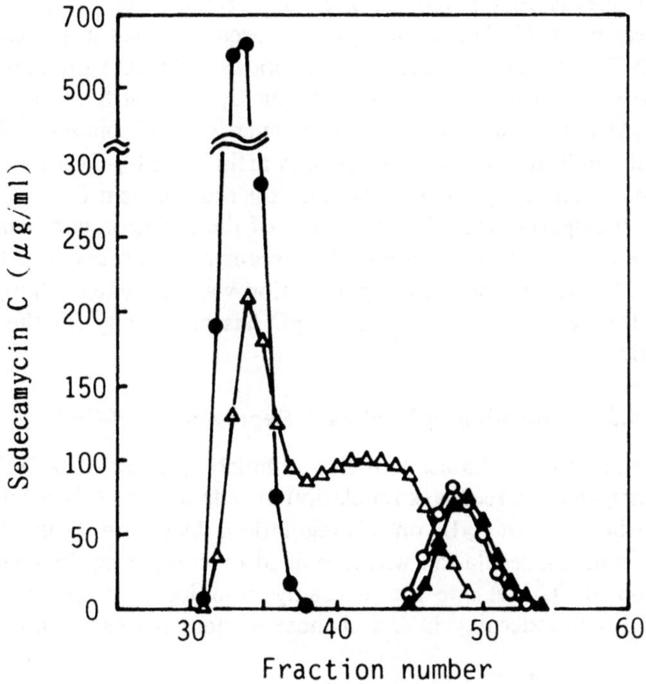

Figure 19. Affinity of α-, β-, and γ-CyD for sedecamycin C (gel filtration). Ten milliliters of a 1.8% aqueous solution of each CyD containing sedecamycin C was passed through a Cellulofine GCL-90-m column (3.1 cm diameter × 52 cm length) and eluted with water at a flow rate of 2.8 ml/min. Each 5-ml fraction was collected and assayed for sedecamycin C: ○: sedecamycin C without CyD; ▲, with α-CyD; ●: with β-CyD; ▽: with γ-CyD. (Sawada et al.[53])

same as that described in Figure 17) with water at a flow rate of 2.8 ml/min. Each 5-ml fraction was collected and assayed for sedecamycin C by HPLC (Figure 19). Sedecamycin C without CyD was eluted at a maximum level in fraction No. 48 (control). When sedecamycin C was previously mixed with α-CyD, it appeared in the same fraction as the control. The chromatographic peak of sedecamycin C moved to fraction No. 34, when it was mixed with β-CyD.

This means that β-CyD is able to form a tight complex with sedecamycin C. The elution pattern became broad in the case of γ-CyD. Harada[50] reported that β- and γ-CyD have the ability to increase the solubility of sedecamycin C, but that ten times more γ-CyD is necessary to obtain the same level of solubility. These results indicate that β-CyD has high affinity and γ-CyD has some affinity, but α-CyD has no affinity for sedecamycin C.

It is apparent that the formation of the inclusion compound is closely related to stimulation of sedecamycin C production. However, in the case of α-CyD, production was stimulated, though to the least extent, even though α-CyD has no affinity for the antibiotic, as shown before.

2.4.4.2. Abrogation of Feed-back Repression by β-CyD

To study the mechanisms of this stimulatory effect of β-CyD, we investigated the feedback regulation of sedecamycin C biosynthesis and the effect of CyDs on this regulation. As Figure 20 illustrates, the feedback regulation was examined by comparing the incorporation of ^{14}C-Met into the sedecamycin molecule in cells preincubated with sedecamycin C and those without sedecamycin C. The

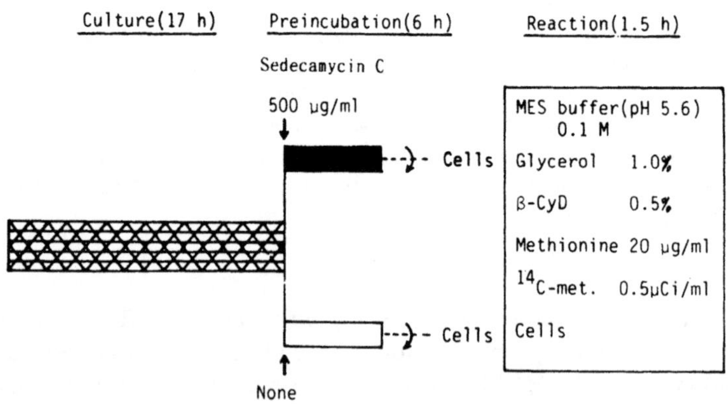

Figure 20. The experiments to study feedback regulation of sedecamycin C biosynthesis: MES: 2-(N-morpholino)ethanesulfonic acid; ^{14}C-met: [^{14}C]methionine. (Sawada et al.[53])

reason cells cultured for 17 h were used is that this phase of cell growth is just before the antibiotic is produced and the cells are not yet affected by the product. After preincubating cells for 6 h in the presence or absence of 500 μg/ml of sedecamycin C, cells were harvested by centrifugation and the incorporation of ^{14}C-Met was measured over 90 min.

A shown in Figure 21, the incorporation activity in cells preincubated with sedecamycin C was distinctly less than that in cells preincubated without the antibiotic. When the experiment was carried out using α-CyD-augmented culture broth, Figure 21(B), the incorporation activity of cells was distinctly higher, and it was more markedly repressed by the preincubation with sedecamycin C. We cannot explain conclusively the reason why the activity of cells cultured in the presence of α-CyD was more strongly repressed by

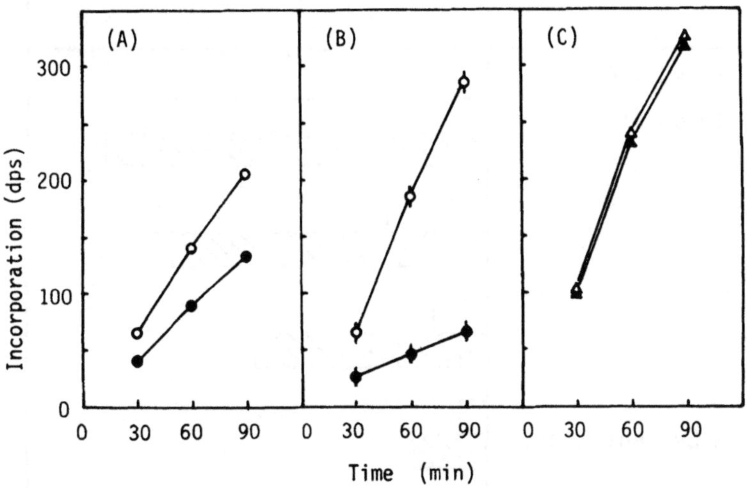

Figure 21. Effects of preincubation of cells with sedecamycin C on [^{14}C]methionine incorporation activity. (A) Cultured in the absence of CyD: ●: preincubated with sedecamycin C, 500 μg/ml; ○: without sedecamycin C. (B) Cultured in the presence of α-CyD: ⧫: preincubated with sedecamycin C; ϕ: without sedecamycin C. (C) Cultured in the presence of β-CyD: ▲: preincubated with sedecamycin C; △: without sedecamycin C. (Sawada et al.[22])

sedecamycin C. In the case of the β-CyD-augmented culture, the incorporation activity was not affected at all by preincubation with sedecamycin C, Figure 21(C). Sedecamycin C, therefore, seems to repress its own biosynthesis, but its activity as a repressor is abrogated by the formation of an inclusion complex with β-CyD.

2.4.4.3. Effect of CyDs on the Cell Surface

In addition to canceling feedback repression, direct interaction of CyD with the cells is thought to be another important mechanism. The stimulatory effect of β-CyD is greatest when it is added at the onset of fermentation (Sawada et al.[51]). The effect of α-CyD is also

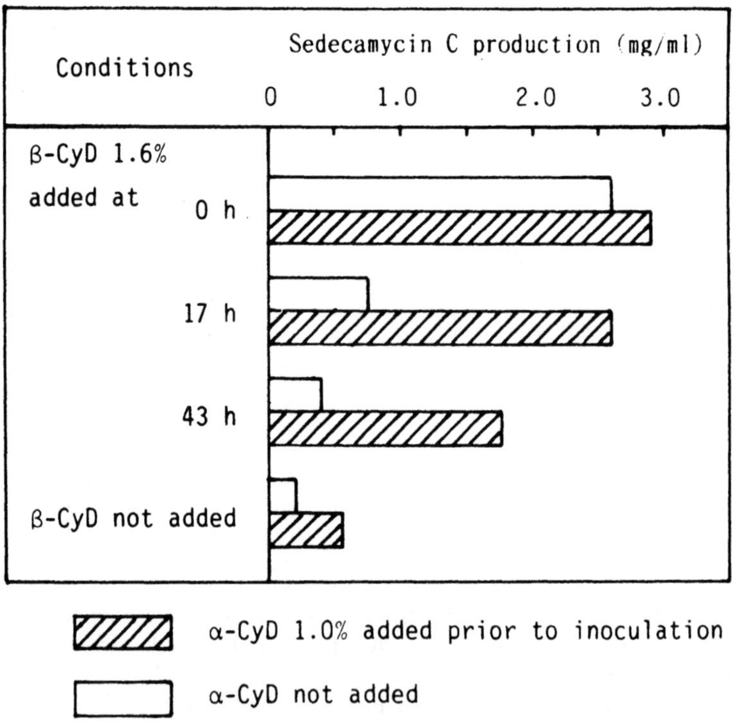

Figure 22. Complementary effect of α- and β-CyD on the stimulatory production of sedecamycin C. (Sawada et al.[53])

noteworthy: α-CyD, which has no affinity for sedecamycin C, stimulates antibiotic production, though the effect is not as significant as that of β-CyD (Table 10, Figure 18). As Figure 22 shows, α-CyD has a complementary effect on the stimulation by β-CyD: When α-CyD is added at the onset of fermentation, β-CyD strongly stimulates production, even if it is added in the production phase. This means that α-CyD or β-CyD is necessary in the growth phase for maximum stimulation.

It was apparent that cells grown in the presence of α-CyD or β-CyD had pulpy mycelia, and the results of an assay of the hydrophobic interaction of the cells with liquid hydrocarbon indicated that the cells had a hydrophilic surface (Rosenberg et al.[55], data not shown). Tendency to show a pulpy morphology and the ability of CyD to change the hydrophobicity of the cell surface were in the order of α-CyD > β-CyD > γ-CyD. Although the relationship between the hydrophobicity of the cell surface and the secretion of sedecamycin C has not been clarified, it is supposed that the change of the cell surface to a hydrophilic condition might result in the promoted secretion of the hydrophobic antibiotic.

Table 11 shows the summary of the effects of CyDs. Based on these results, the following mechanism has been proposed: Two mechanisms are involved in the stimulatory effect of CyD: one is the abrogation of feedback repression and the other is the modification of the cell surface to promote secretion of the hydrophobic antibiotic. This hypothesis is supported by the complementary effects of α-CyD and β-CyD, as shown in Figure 22.

Table 11. Summary of the effect of α-CyD, β-CyD, and γ-CyD.

	α-CyD	β-CyD	γ-CyD
I] Formation of inclusion compound with sedecamycin C (masking of repressor)	−	+++	++
II] Effect on cell surface of the microorganism	+++	++	±
→ Stimulation of sedecamycin C production	+	++++	++

Source: Sawada et al.[6]

2.5. Development of Industrial Process

2.5.1. The Improvement of Strain and Media

2.5.1.1. Breeding of the Strain Producing No By-products

The original strain, *Streptomyces rochei var. volubilis* IFO 12507, produces lankamycin and T-2636 M as by-products in addition to sedecamycin. Lankamycin is a 14-membered macrocyclic antibiotic, and T-2636 M is a polyene antibiotic (Harada et al.[56]).

Various kinds of mutants were derived from the original strain by UV irradiation. After they were cultured in seed media, they were transferred to 20 ml of fermentation medium consisting of 10% glycerol, 2% proflo, 1% polypepton, 0.5% corn steep liquor, 0.5% NaCl, 0.1% $FeSO_4 \cdot 7H_20$, 0.0025%, $CuSO_4 \cdot 5H_2O$, 2% β-CyD, and 0.01% Actcol (antifoaming agent), pH 5.8 (GP-medium) in 200-ml Erlenmeyer flasks described before (Sec. 2.4.1). The production of sedecamycin C was measured by HPLC as described in Figure 16.

Lankamycin can be assayed by TLC using a chromatoscanner, a silica-gel plate (60F-254, Merck Co.), and a solvent system (Chloroform: Methanol = 93: 7), and T-2636 M can be assayed by a biossay system using *Rhodotorula rubra* IFO 907. Among the mutants, three appropriate mutants were found. Mutant strain mu-1 produced sedecamycin C only, mu-2 lankamycin only, and mu-3 T-2636M only (Sawada et al.[53]). Table 12 summarizes the results.

Table 12. Isolation of mutants producing sedecamycin C, lankamycin, or T-2636M.

Strain[a]	Antibiotic[b] productivity		
	Sedecamycin C (μ g/ml)	Lankamycin (μ g/ml)	T-2636M (μ g/ml)
Parent	2100	460	240
mu-1	2100	trace	trace
mu-2	trace	400	trace
mu-3	trace	trace	650

[a] *Streptomyces rochei* var. *volubilis* IFO 12507 and its mutants.
[b] The cultures were carried out for 4 days and the antibiotic productivity was assayed using GP medium as described in the text.

2.5.1.2. Derivation of an HON-Resistant Strain

Biosynthesis of sedecamycin was studied (Uramoto et al.[59], Hatano et al.[60]), and it was assumed that glycine is a precursor related to the structure of the peptide bond of the antibiotic and that the methyl group of methionine is incorporated into the methyl-group residues of sedecamycin. Considering that amino acids play an important role in the biosynthesis of an antibiotic, it occurred to us that mutation of amino acid–biosynthesis regulation might have an effect on the production of sedecamycin.

We tried to derive amino acid–resistant mutants from *Streptomyces rochei var. volubilis* RS-50 (mu-1 type) to get a mutant strain having high sedecamycin productivity. Among the many kinds of amino acid analogues, four analogues that repressed the growth of strain RS-50 in the concentration of 100 μg/ml were found. They are glycine hydroxamate, DL-serine hydroxamate, DL-δ-hydroxy-γoxo-norvaline (HON), and DL-5-fluorotryptophan. After harvesting spores of the strain RS-50, the spores were UV irradiated. They were then spread on agar medium containing an amino acid analogue at the concentration of 100 μg/ml. The agar medium consisted of 1% soluble starch, 0.2% $(NH_4)_2SO_4$, 0.1% K_2HPO_4, 0.1% $MgCl_2$, 0.1% NaCl, 0.0001% $FeSO_4 \cdot 7H_2O$, 0.0001% $MnCl_2 \cdot 4H_2O$, 0.0001% $ZnSO_4 \cdot 7H_20$, 0.2% $CaCO_3$, and 2% agar, pH 7. After incubation at 28°C for 10 days. Colonies growing on the agar medium were isolated and transplanted to a slant medium. The mutants thus obtained were cultured under the

Table 13. HON-resistant mutants and their sedecamycin C productivity.

Strain	Sensitivity to HON[a] MIC (μ g/ml)[c]	Sensitivity to 2-deoxy-glucose MIC (μ g/ml)[c]	Sedecamycin c productivity[b] (μ g/ml)
RS-50 (mu-1)	50	100	2100
HON^R-190	100	150	2800
HON^R-111	150	200	3500
HON^R-74	190	280	4700

[a] DL-δ-hydroxy-γ-oxo-norvaline.
[b] After 4 days of culturing in GP medium, sedecamycin produced was assayed by HPLC as described in the text.
[c] Minimum inhibitory concentration.

same conditions as described before (Sec. 2.5.1.1), and the sedecamycin production tests were conducted. As a result, no correlations were found, except in the case of HON, between the increase in the resistance to amino acid analogues and the production of sedecamycin. Only the HON-resistant mutants were found to have higher sedecamycin productivity.

As shown in Table 13, an increase in the minimum inhibitory concentration of HON is found to correlate with increased sedecamycin productivity. For example, HON^R-74 having the highest resistance to HON showed the highest sedecamycin productivity.

Furthermore, as the table shows, these mutants showed resistance to not only HON, but also 2-deoxyglucose. The mechanisms of these correlations are still unclear.

2.5.1.3. Derivation of 2-Deoxyglucose-Resistant Mutants

As there were some correlations between sedecamycin productivity and the resistance to deoxyglucose, it was expected that an increase of resistance to 2-deoxyglucose would affect sedecamycin productivity. Therefore, we tried to derive a strain completely resistant to 2-deoxyglucose and obtained strain DG^R-20-1, which could grow on medium containing more than 1% of 2-deoxyglucose. Though there were no differences in productivity between the mutant and the parent strain in the test using GP medium, DG^R-20-1 showed much higher sedecamycin production using a medium consisting of 10% dextrin, 0.5% proflo (Trader Oil Mill Co., USA), 2.0% defatted soybean flour, 1% corn steep liquor, 1.3% corn gluten meal, 0.025% p-aminobenzoic acid, 0.2% $(NH_4)_2SO_4$, 0.005%

Table 14. 2-Deoxyglucose-resistant mutants and sedecamycin C and carboxyl esterase productivity using GP and DS media.

Strain	Sedecamycin C (μg/ml)[a]		Carboxyl esterase (U/ml)[b]	
	GP medium	DS medium	GP medium	DS medium
HON^R-74	4700	500	260	62
DG^R-20-1	4740	4800	501	505

[a] After culturing for 4 days, sedecamycin C was assayed by HPLC as described in the text.
[b] After culturing for 4 days, carboxyl esterase was assayed by the method described in Sec. 2.5.2.1 using 4-nitrophenol acetate.

$CuSO_4 \cdot 5H_2O$, 0.3% $NiSO_4 \cdot 7H_2O$, 2% β-CyD, and 0.01% Act-col (DS medium).

DS medium is more suitable for industrial use than GP medium, because the cost of the materials for DS medium is lower than that for GP medium. From the point of view of developing an industrial process, strain DG^R-20-1 is more appropriate than strain HON^R-74. In addition, the carboxyl esterase activity in DG^R-20-1 was found to be about twice that in HON^R-74.

These results are summarized in Table 14.

2.5.2. Studies on Acetylation of Sedecamycin C

2.5.2.1. Preparation of Estarase

Carboxyl esterase (EC 3.1.1.1.) is produced and excreted into the culture broth by sedecamycin-producing *Streptomyces*. The enzyme catalyzes the selective acetylation of the hydroxyl group at the 13-position of sedecamycin C to give sedecamycin A specifically.

In the industrial process, the purification of the enzyme is not always necessary because the filtrates of culture broth contain strong esterase activity that easily esterifies sedecamycin C in the presence of an acetyl group donor. However, basic studies using purified esterase are useful to eludicate the mechanisms of acetylation and to search for the optimal conditions for the reaction.

The method of esterase preparation is as follows.

A 100-liter portion of GP medium (Sec. 2.5.1.1) was placed in a 200-liter fermentor, and then it was sterilized by steam at 120°C for 20 min.

Five liters of the seed broth cultured using *Streptomyces rochei var. volubilis* HON^R-190 was transplanted into this culture medium and incubated at 24°C for 96 h with an aeration of 1 VVM and agitation at 165 rpm. Twenty liters of water and 2 kg of HyfloSuper-Cel (Manville Products Co., USA) was added to 60 liters of the culture broth thus obtained, and the mixture was filtered to give 70 liters of the filtrate.

Ten liters of the filtrate was placed in a 60-liter vessel, and 40 liters of ethanol was added. The mixture was then stirred well with a stirring rod and allowed to stand at 5°C for 12 h so that proteins would precipitate. The supernatant liquid was removed with a siphon to obtain a white, turbid precipitate.

The precipitate was centrifugated at 2000 *g*, keeping the temperature at 5°C to reduce the water content of the precipitate as much as possible. The obtained product was washed with ethanol and again centrifuged by the same procedure as before. The precipitate was collected and dried under reduced pressure (50 mm Hg) at 10°C for 24 h to obtain about 250 g of powder.

The powder was dissolved in 500 ml of 0.05 M tris(hydroxymethyl)aminomethane-hydrochloric acid buffer (pH 7.4), and insoluble matter was removed by centrifugation of 2000 *g* at 50°C. After 75 g of ammonium sulfate was added to the solution and dissolved, the solution was subjected to interaction chromatography using a column (diameter: 40 mm; length: 300 mm) packed with Butyl Toyopearl 650C (TOSOH Co.) equilibrated with the same buffer containing 15% ammonium sulfate. It was then eluted with the same buffer containing ammonium sulfate (3%).

By a method to assay carboxyl esterase activity using 4-nitrophenyl acetate (Huggins and Lapides[61]), active portions were identified and 400 ml of carboxyl esterase fractions were collected.

After 65 g of ammonium sulfate were added to the solution and dissolved, the solution was subjected to a column (diameter: 30 mm; length: 250 mm) packed with Toyopearl HW 65C (Interaction Chromatography, TOSOH Co.); the elution was conducted with the same buffer containing ammonium sulfate (3%). The fractions showing carboxyl esterase activity were collected and subjected again to the Butyl Toyopearl (the same as shown before) under the same conditions described before.

Five hundred milliliters of active fractions eluted from the Butyl Toyopearl column were collected. The resulting solution was stirred with a stirrer in a 3-liter Erlenmeyer flask, and 250 g of ammonium sulfate were added gradually to precipitate the protein. The mixture was allowed by stand at 5°C for 24 h, and precipitate was collected by centrifuging at 2000 g at 5°C.

The precipitate was dissolved in 100 ml of 0.05 M tris-(hydroxylmethyl)aminomethane-hydrochloride buffer (pH 7.4). The solution was subjected to gel filtration using a column (diameter: 100 mm; length: 1500 mm) packed with Sephadex G-100, and the elution was

Table 15. Summary of the esterase purification procedure.

Step	Total protein (mg)	Total activity (U)	Specific activity (U/mg protein)	Yield (%)
Extract	210,000	2,520,000	12	100
Ethanol precipitate	63,200	2,300,000	36	91
Butyl Toyopearl(I)	10,800	2,200,000	204	87
Toyopearl HW	3,000	1,340,000	447	53
Butyl Toyopearl(II)	430	1,150,000	2,680	46
Sephadex G-100	380	1,030,000	2,700	41
DEAE Sephafex A-50	250	688,000	2,750	27

conducted with the same buffer solution as before at a flow rate of 10 ml per 5 min.

The fractions showing carboxyl esterase activity were collected and were subjected to linear gradient ion-exchange chromatography using an increasing salt concentration and a column (diameter: 50 mm; length: 400 mm) packed with DEAE Sephadex A-50. The elution was conducted with 2 liters of the same buffer containing 1 M NaCl in the first vessel and 2 liters of the same buffer without NaCl in the second vessel.

Three hundred milliliters of active fractions was collected. After being concentrated with a protein concentrator, it was injected into a tube made of cellophane for dialysis and placed under 2 liters of water in a 3-liter Erlenmeyer flask to conduct dialysis. The dialysis was continued for 24 h by stirring the external water with a magnetic stirrer. After completion of dialysis, the solution in a tube was put in a 2-liter eggplant-type flask and lyophilized under reduced pressure (20 mm Hg) for 24 h using a lyophilizer to obtain powdery carboxyl esterase, which originated in *Streptomyces rochei* var. volubilis. The obtained carboxyl esterase was confirmed to be single protein by SDS gel electrophoresis, and the molecular weight was determined to be about 30,000. The purification process is summarized in Table 15.

2.5.2.2. Enzymatic Acetylation of Sedecamycin C Using Acetic Anhydride in the Presence of Esterase

Carboxyl esterase (EC 3.1.1.1.) is produced and excreted into the culture broth by sedecamycin-producing *Strepyomyces*. Sedecamycin A is produced from sedecamycin C, which has hydroxy substituents at the 7 and 13 position. The hydroxy group at the 13 position is specifically converted into an acetyl group by the esterase (refer to Figure 14). In the course of our studies on the esterification, we found that acetic anhydride could be used much more efficiently than ethyl acetate or acetic acid as the acetyl-group donor.

In order to determine the Michaelis constant and maximal reaction velocity, we established a one-phase reaction system using the solvent-soluble esterase modified with activated polyethylene glycol. We prepared the modified esterase, according to the improved procedure of Matsushima,[62] which was based on the original method.[63] A 50-mg portion of purified carboxyl esterase preparation was dissolved in 8 ml of 0.1 M sodium borate buffer (pH 9.5) and 1 g of 2,4-bis(o-methoxy-polyethylene glycol)-6-chloro-s-triazine (molecular weight of polyethylene glycol: 6000) was added, followed by stirring at 5°C for 1 h. To stop the reaction, 72 ml of 0.2 M potassium phosphate buffer (pH 7.2) was added to the reaction mixture.

The reaction mixture was subjected to ultrafiltration (ultrafiltrator: 8050 type; Amicon Division of Grace Corp., USA) with milliporefilter YM-10 and washed with 400 ml of 0.2 M potassium phosphate buffer (pH 7.0) to remove unreacted 2,4-bis(o-methoxy-polyethylene glycol)-6-chloro-s-triazine. Then, the solution containing enzyme modified with polyethylene glycol was put in a 500-ml eggplant-type flask and lyophilized under a pressure of 20 mm Hg for 24 h to obtain 600 mg of carboxyl esterase modified with polyethylene glycol (PEG-modified esterase). The PEG-modified esterase can be dissolved in methyl isobutyl ketone.

Two milliliters of an 11 mM sedecamycin C and 50 mM acetic anhydride solution in methyl isobutyl ketone was put in a 20-ml test tube with a stopper. A PEG-modified esterase solution, which contained 3.6 mg of modified enzyme in 0.3 ml of 0.2 M tris(hydroxyl)aminomethane-maleic acid buffer (pH 7.0), was added to the tube. The tube was then placed in a shaking apparatus and shaken (80 spm) at 37°C.

By using the HPLC method described in Figure 16, the amount of sedecamycin A formed was measured.

As a control, the same procedures as before were conducted

Table 16. Comparison of acetyl-group donors in the acetylating reaction.

Reaction time (min.)	Acetyl-group donor		
	Acetic anhydridec (50 mM)	Ethyl acetate (300 mM)	Acetic acid (100 mM)
0	–[a]	–	–
10	5.58	0.54	–
20	8.11	0.69	–
30	9.09	0.95	–
45	9.30	1.45	–
80	10.83	1.98	–
110	10.83	2.10	–

[a] Not more than 0.1 mM.

using ethyl acetate (300 mM) or acetic acid (100 mM) instead of acetic anhydride (50 mM). The results are shown in Table 16. The table clearly indicates that acetic anhydride is superior to ethyl acetate or acetic acid in terms of the rate of sedecamycin A formation.

The results of changing the concentration of the acetyl-group donor to those shown in Table 17, with the reaction time being 20 min., are shown in the table. When using acetic acid as the

Table 17. Acetylation reaction velocity and the concentration of the acetyl-group donor.

Acetyl-group donor	Acetic anhydride		Ethyl acetate	
	$[S]^a$	v^b	$[S]$	v
	1	0.55	500	0.85
	2	1.06	1,000	1.05
	3	1.40	2,000	1.76
	5	2.51	3,000	2.39
	7	3.44	4,000	3.28
	10	4.76	5,000	3.95
	15	6.93	6,000	4.33
	20	7.28	7,500	4.95
			10,000	6.13

[a] Concentration of substrate (acetic anhydride or ethyl acetate) (mM)
[b] Reaction velocity (mM/20 min.)

acetyl-group donor, the formation of sedecamycin A could not be detected.

By using Lineweaver-Burk's equation,[64]

$$1/v = (K_m/V_{max}) \cdot (1/[S]) + 1/V_{max}$$

where v = reaction velocity
V_{max} = maximal reaction velocity
$[S]$ = concentration of substrate
K_m = Michaelis constant

the reaction using acetic anhydride as the acetyl group donor was compared with that using ethyl acetate.

By using the values of $[S]$ and v shown in Table 17, the values of K_m and V_{max} were calculated by means of the least-squares method. The results are shown in Table 18.

Table 18 clearly indicates that when acetic anhydride is used as the acetyl-group donor, the value of K_m is much smaller than when ethylacetate is used as the acetyl-group donor.

Thus, acetic anhydride was chosen as the acetyl-group donor in the industrial process. Various kinds of organic solvents were also examined in addition to methyl isobutyl ketone. Ethyl acetate was chosen for the industrial process, because the solvent has some advantages: it costs less and sedecamycin A is highly soluble in it.

2.5.3. Process for the Industrial Production of Sedecamycin

2.5.3.1. Fermentation Process

One hundred liters of DS medium was placed in a 200-liter fermenter and subjected to steam sterilization at 120°C for 20 min. Five liters

Table 18. Comparison of K_m and V_{max} values when using acetic anhydride and ethyl acetate.

Acetyl-group donor	K_m^a	V_{max}^b
Acetic anhydride	7.69	6.59
Ethyl acetate	3010	5.40

[a] Michaelis constant (mM).
[b] Maximal reaction velocity (mM/20 min.).

of the same seed medium as described before (Sec. 2.5.2.1) was transplanted into the culture medium and incubated at 24°C for 5 days under aeration of 1 VVM and agitation at 165 rpm.

Thirty-five liters of water and 3 kg of Hyflo Super-Cel (Manville Products Co., USA) were added to the culture broth (102 liters) thus obtained and the mixture was filtered to give 120 liters of filtrate.

By using the HPLC method described previously (Figure 16), the content of sedecamycin C was measured. Sedecamycin C was found to be present in a concentration of 4100 μg/ml of the filtrate, and the filtrate was found to have strong carboxyl esterase activity by the method described previously (Sec. 2.5.2.1).

2.5.3.2. Acetylation and Purification Process

The flow diagram of the acetylation and purification process is shown in Figure 23.

One hundred liters of ethyl acetate containing acetic anhydride at the concentration of 100 mM was added to a 500-liter reactor equipped with a stirrer. The previous 120 liters of the filtrate was added to the reactor. The reaction was carried out for 40 min. at 37°C with stirring.

After it was confirmed by the HPLC method that the sedecamycin C had been completely acetylated and converted to sedecamycin A, the mixture was put into an extractor to remove the organic solvent phase that contained sedecamycin A.

After the obtained organic solution was dried over Na_2SO_4 for 2 h, it was placed in a vacuum still and concentrated under reduced pressure to dryness and 450 g of sedecamycin A as a crude powder was obtained.

This powder was dissolved in 50 liters of chloroform, and the solution was applied to a silica-gel (Merck Co.) column (diameter: 300 mm; length: 700 mm) equilibrated with chloroform. After washing with 50 liters of chloroform, the elution was conducted with a mixture of ethyl acetate and chloroform (ethyl acetate: chloroform = 3: 7 *v*/*v*). Fractions of 500 ml each were monitored by ultraviolet spectroscopy and the fractions containing sedecamycin A were collected.

The pooled sedecamycin A-containing fractions were concen-

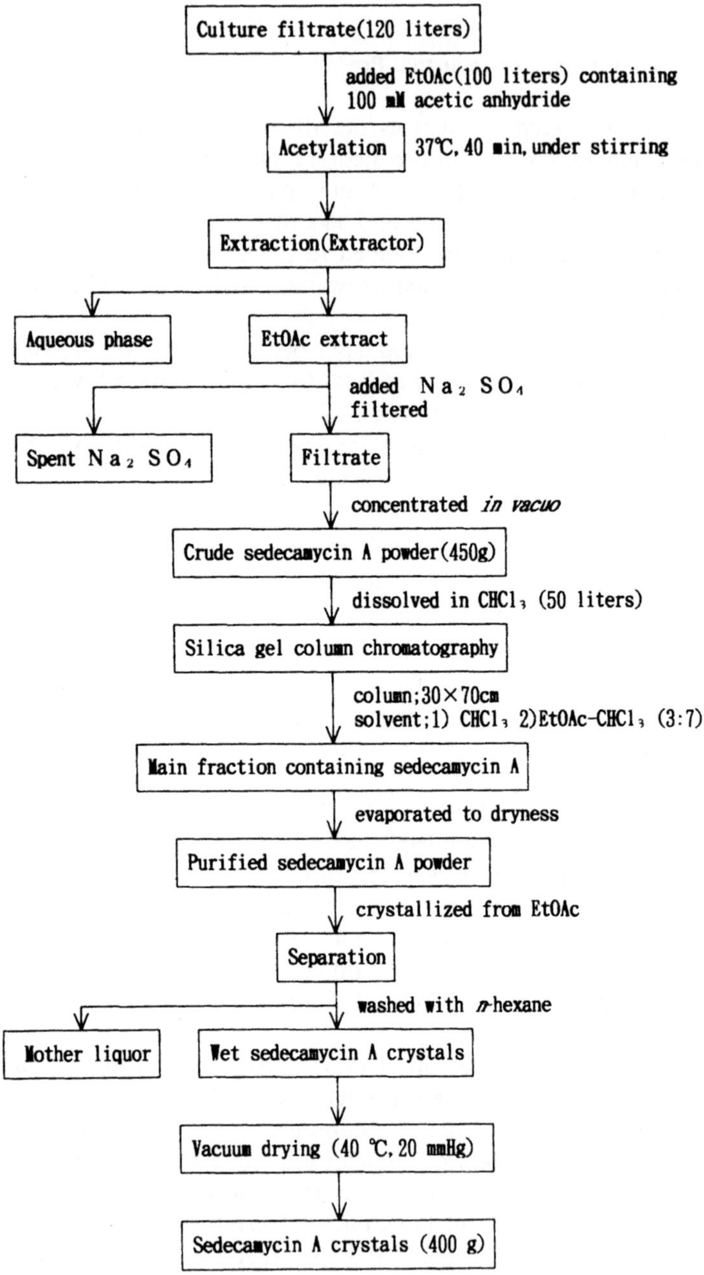

Figure 23. Process flow diagram for producing sedecamycin A.

trated under reduced pressure to dryness using the same method as described before to give sedecamycin A powder.

Forty liters of ethyl acetate were added to the powder, and the resultant solution was concentrated under reduced pressure to 1.6 liters and then cooled to 5°C. Crystals of sedecamycin A that precipitated out were collected, washed with n-hexane, and dried under a pressure of 20 mm Hg at 40°C to obtain 400 g of sedecamycin A as crystals.

CHAPTER 3

Studies on the Production of Bialaphos from a Biochemical Engineering Viewpoint

Hidehi Takebe and Hideo Tanaka

3.1. Introduction

Success in the industrial production of microorganism metabolites depends mainly not only on the productivity of the producing strain used, but also on its characteristics. In order to select strains suitable for the industrial production of microorganism metabolites at a minimum cost, an improvement in productivity is, of course, extremely important. At the same time, many problems that exist in the entire process, starting with raw materials to the final product, must be considered. Problems to be considered in the breeding of producing strains include the possibility of using an inexpensive substrate, high productivity under low oxygen supply, reproducibility of high productivity, less by-product, the ease of separation of cells from the fermentation broth, and easy purification of the product. Although there have been a number of reports on breeding studies focusing on how to obtain high productivity strains, few reports can be found that discuss these matters from a biochemical engineering viewpoint at the breeding study stage and that overview all problems involved in the whole process from culture to downstream treatment.

Recently, Meiji Seika Kaisha, Ltd., has succeeded in the industrial-scale production of bialaphos, a herbicide, by employing a microbe culture method. Bialaphos is a tripeptide, consisting of 2-amino-4 (hydroxy)(methyl) phosphinoyl butyric acid (AMPB) and two alanine residues (Ogawa et al. *Sci. Reports of Meiji Seika Kaisha*, **13**, 42 [1973]). The mode of action is known to be inhibition of glutamine synthetase[65]. It is easily degraded by soil microorganisms so that no environmental contamination is expected. It exhibits excellent herbicidal activity. In the industrial production of bialaphos, there are various biochemical engineering problems to be overcome before commercialization. Our success in industrial-scale production of bialaphos comes from breeding and biosynthesis studies,[66-68] in which the various problems described before are discussed by comparing a high-productivity strain selected from a biochemical engineering viewpoint with a low-productivity strain.

3.2. Breeding of Bialaphos-Producing Strain

3.2.1. Characteristics of Wild Strain

Prior to breeding the producing strains, various problems that could potentially arise in the parent wild strain were investigated. Figure 24 shows the time course of bialaphos fermentation in liquid culture (medium with glycerol) using a wild strain (*Streptomyces hygroscopicus* ATCC 21705). First, the production of bialaphos by the wild strain was possible only when an expensive glycerol-based medium was used. When it was replaced with glucose (not shown in the figure), the production activity of bialaphos virtually disappeared. Even in glycerol, the bialaphos concentration was very low with high glycerol consumption and high cell concentration (dry cell weight as high as 70–80 mg/ml).

This proves that the strain was affected by typical carbon catabolite regulation, causing low substrate efficiency for product accumulation. Second, the high cell concentration not only required a high oxygen supply in liquid culture, but it made it difficult to recover the filtrate from the cultivation broth due to its high viscosity. Furthermore, bialaphos once formed was decomposed to AMPB, a by-product of bialaphos lacking two alanines, causing AMPB accumulation in the medium. This occurred in the later period of cultivation during which glycerol disappeared and the pH increased.

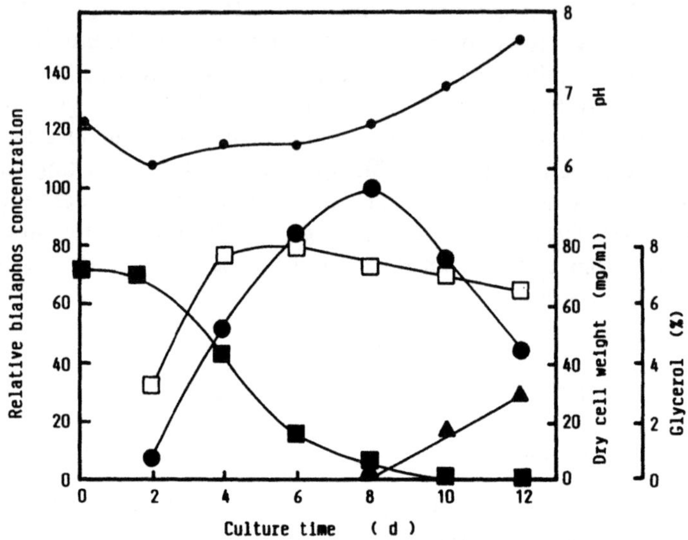

Figure 24. Time course of bialaphos fermentation by the wild strain in P-1 medium with glycerol. ●: bialaphos; ▲: AMPB; ●: pH; □: dry-cell weight; ■: glycerol P-1 medium, which consisted of 7.0% glycerol, 4.4% Bactosoyton, 0.33% KH_2PO_4, 0.085% Na_2HPO_4, 1.15% TES, and 0.0001% $CoCl_2 \cdot 2H_2O$ (pH 7.0) was used. Thirty ml of the medium were placed into a 250-ml flask and cultivation was carried out by inoculating 2% of the precultured mycelia in S-1 medium containing 2.0% soluble starch, 1.0% Polypepton, 0.3% meat extract, and 0.05% KH_2PO_4 (pH 7.0). The flask was shaken on a rotary shaker (220 rpm, 70-mm stroke) at 28°C.

3.2.2. Target Factors in the Breeeding Study

Target factors in breeding studies that may be selected from a biochemical engineering viewpoint are summarized in Table 19.

The study was initiated based on 4 such target factors: (a) small colony type strains, (b) strains free from catabolite regulation of carbon and nitrogen sources, (c) strains having low decomposition activities, and (d) strains with high product concentrations under conditions of low oxygen supply (mild aeration–agitation).

On the basis of these target factors, strains suitable for the indus-

Table 19. Targets of breeding study from a biochemical engineering viewpoint and their expectations.

Target	Expectations
Small colony type strains	1. High productivity per unit cell mass 2. Efficient utilization of substrate 3. High filtration rate and yield of cells from fermentation broth 4. Low oxygen requirement
Strains free from catabolite regulation of carbon and nitrogen sources	1. Inexpensive substrate utilization (medium cost reduction) 2. Efficient utilization of substrátes (medium cost reduction)
Strains with low decomposition activities	1. Increase in productivity 2. Stable productivity 3. Easy determination of termination point in culture process
Strains exhibiting high productivity even under mild aeration and agitation in liquid culture	1. Less oxygen demand 2. Less foam 3. Easy control of culture 4. Maximizing yields by increasing the liquid volume

trial production of bialaphos were selected by various methods outlined in what follows.

3.2.2.1. Selection of Small Colony Type Strains

Small colony type strains are preferred because they have properties such as high product concentration per unit cell, high filtration efficiency, and low oxygen requirement. A selection study was performed in order to obtain strains having small colonies and large inhibition zones by the agar piece method (*Proteus* sp. MB-838 was used as a detector of bialaphos formation). As shown in Figure 25, with the progression of breeding (from right to left), colonies tended to become small, lose the ability to form spores, and become wet.

3.2.2.2. Selection of Catabolite Regulation-Free Strains

Strains free from catabolite regulation of carbon sources are preferred because they have a high potential to reduce medium costs.

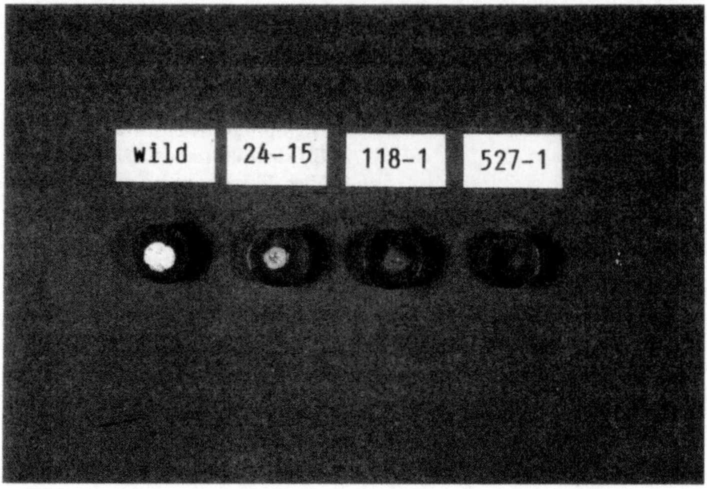

Figure 25. Colony types of strains with progress in breeding (from right to left). As breeding progressed, colonies tended to become smaller.

Accordingly, glucose-based agar medium was prepared by switching glycerol in the medium as a carbon source. The selection study was performed in order to obtain strains having small colonies and large inhibition zones using a glucose medium.

As shown in Table 20, a good breed of strain (strain 118-1) was

Table 20. Effects of carbon sources on relative bialaphos concentrations of the low-productivity (24.15) and high-productivity (118.1) strains[a].

Carbon source	Relative bialaphos concentration	
	Low-productivity strain (24–15)	High-productivity strain[b] (118–1)
Glycerol	100	190
Glucose	30	510
Starch	60	450
Sucrose	Trace	300
Maltose	Trace	470
Lactose	Trace	180

[a] When a glucose-based medium was used, the dry cell weight of the low-productivity strain was 62 mg/ml and that of the high-productivity strain was 43 mg/ml.
[b] Free from glucose catabolite regulation.

obtained. This strain exhibits a remarkable decrease in the amount of cells and an increase in bialaphos concentration in liquid culture of glucose-based medium, as well as less carbon catabolite regulation. This means that strain 118–1 obtained through the procedure made available a change in material from expensive glycerol to inexpensive glucose. The strain produces a concentration 17 times more per unit liquid volume and 25 times more per unit cell mass than the low productive strain 24–15. Similar results were obtained when material was switched from glucose to starch, sucrose, or maltose.

3.2.2.3. Selection of Strains Having Low Bialaphos Decomposition Activity

As shown in Figure 24, in the latter period of cultivation during which glycerol or glucose disappears and pH increases in liquid culture, the concentration of bialaphos tends to decrease. Instead, the concentration of AMPB, a decomposition product, increases in the medium. To eliminate such an unfavorable phenomenon, some approaches may yield good results; for instance, improvements in the medium composition and cultivation conditions (pH control, aeration, and agitation). However, the fundamental solution should be an improvement of the strain itself. Therefore, we attempted to select strains having low bialaphos decomposition activity.

An attempt was made to clarify the optimal conditions for the bialaphos producing strains to decompose the bialaphos once it is formed. As a result, the addition of calcium chloride was found to be the best promoter of bialaphos decomposition. The addition of glutamine, wheat germ, and soluble vegetable protein (Sungrain) as well as high-temperature conditions in the culture also assisted with the rapid decomposition of bialaphos. Thus, the colonies that appeared on A-1 agar medium after mutagenesis were replicated to agar pieces of AP-2 medium composed of 0.5% wheat germ, 0.5% soluble vegetable protein, 0.33% glutamine, 200 μg/ml bialaphos, 0.1% $CaCl_2 \cdot 2H_2O$, and 2.0% agar (pH 7.0), and then cultured for 13 days at 32°C. The agar pieces after cultivation were placed on A-2 medium with *Proteus sp* MB-838, which was used as a detector, and cultured for 16 h at 28°C. The amount of bialaphos was estimated by determining the size of the inhibition zone. Through this procedure, the strains that showed larger zones of bialaphos were selected as low decomposition activity mutants.

In order to confirm whether the strains with superior productivity obtained from this approach remained intact in the liquid culture stage, the following study was carried out. The results for strain 251–2 (improved strain) cultivation broth in liquid culture are shown in Figure 26. No formation of AMPB was recorded even after an 8-day culture period, showing that the decomposition of bialaphos is well controlled. This is also significant when comparing it to the parent strain in which the formation of bialaphos reached a maximum 8 days after starting, AMPB increased late. It is interesting to further note that in addition to controlling the decomposition activity, the glucose consumption rate decreased and the concentration of bialaphos increased by approximately 130–140%.

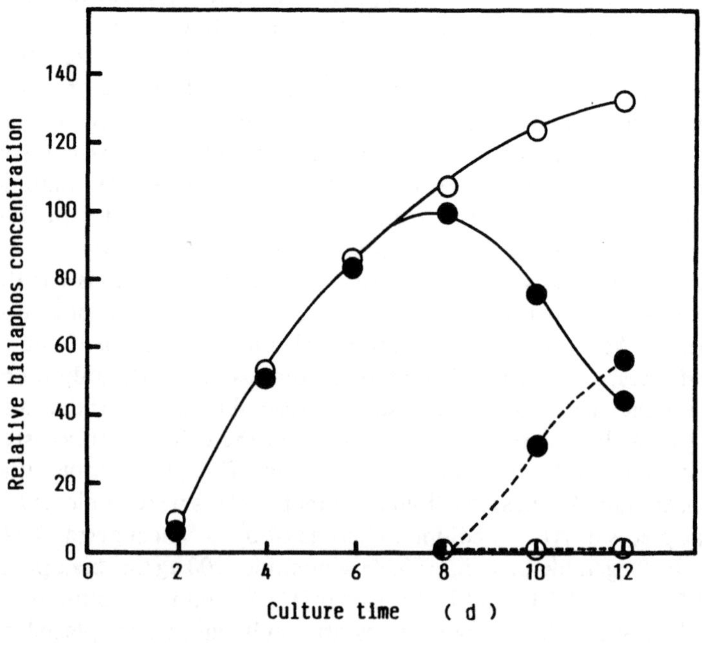

Figure 26. Comparison of bialaphos decomposition between an (○) improved strain 251–2 and (●) the parent strain. —— bialaphos; – – – AMPB.

Based upon these findings, this strain is considered to be a mutant that either lacks the enzyme activity involving bialaphos decomposition or somehow controls the decomposition activity.

3.2.2.4. Selection of Suitable Strains for Industrial production in the Liquid Culture system

The strains obtained by repeating or combining the selection procedure of the previously mentioned agar culture systems were subjected to a liquid culture system under mild shaking conditions in order to select more suitable strains for industrial production under a low oxygen supply. In flask culture, high-productivity strains were selected under the condition of a large volume of meduim. At the same time, the characteristics of broth filtration efficiency and the degree of clearness of the filtrate were used for selection judgment.

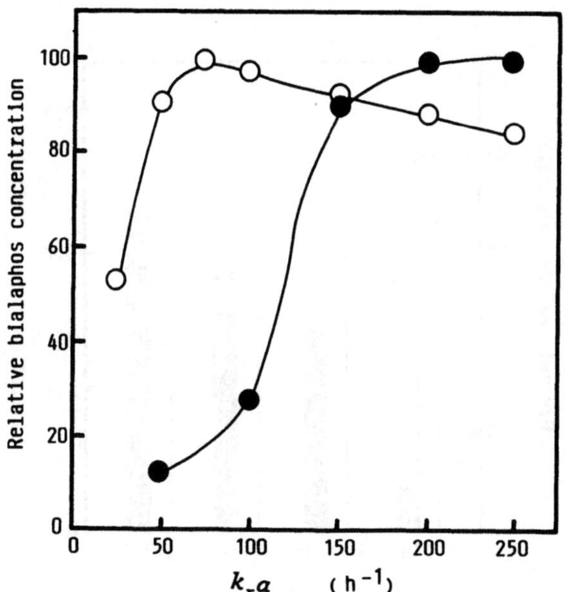

Figure 27. Effects of k_La on the bialaphos concentration in (●) low-productivity strains 24–15 and (○) high-productivity strain 527–1 in P-1 medium with glucose.

Figure 27 shows the results of flask culture of strain 527–1 obtained by repeating and combining the previous selection procedures. This indicates that while in the low-productivity strain culture (strain 24–15), maximum bialaphos product concentration appeared at 200 h^{-1} of volumetric oxygen transfer coefficient ($k_L a$), in strain 527–1 the maximum appeared at as low as 50 to 100 h^{-1}. An interesting fact observed in the culture of the high-productivity strain was that as the oxygen supply increased, the product concentration decreased. This may be attributed to the activation of primary metabolism under high oxygen pressure, resulting in a decrease in secondary metabolism, or the suppression of enzymes related to secondary metabolism for bialaphos biosynthesis by oxygen. Figure 28 shows the dry cell weight and the rate of filtration through filter paper when the successive strains obtained during breeding study were cultured in flasks. This demonstrates that as breeding proceeded, the cell concentration decreased markedly; conversely, the

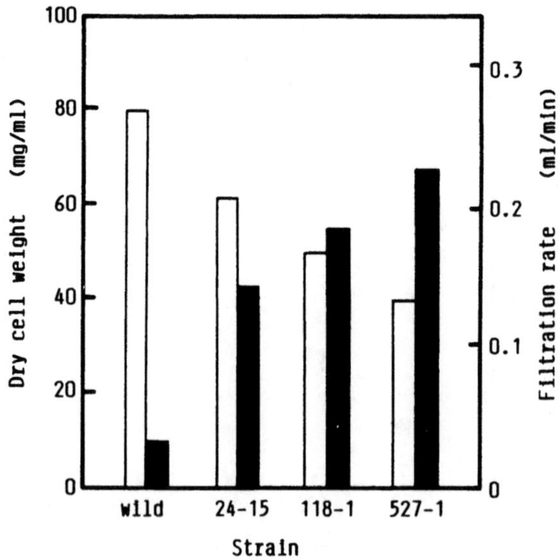

Figure 28. Comparison of the wild strain with the improved strain with respect to filtration rate in P-1 medium with glucose. □: dry-cell weight; ■, filtration rate.

rate of filtration becomes high. It was found that for small colony type strains, such as 527–1, it was possible to achieve highly efficient filtration and a low oxygen requirement.

3.2.3. Overall assessment of Pilot Plant Scale Production by a Selected High-productivity Strain

All tests performed previously were carried out in a flask, using synthetic medium. A scale-up test was then performed in 600-liter tank using 527–1 and the wild strain. The medium was based on commercially available natural organic substances, that is, P-2 medium composed of 4.4% glucose, 3.5% wheat germ, 2.25% soluble vegetable protein, 0.3% KH_2PO_4, and 0.0001% $CaCl_2 \cdot 2H_20$ (pH 6.5). The volume of the medium was 300 liters. The air flow rate for aeration was 300 liters per minute and the agitation speed was 200 to 300 rpm. The seed culture, prepared as before, was inoculated at a dose corresponding to 2% of the medium. Filtration was carried out using a filter press (Noritake filter press; filtration area 7.8m^2), with filtration aids that had previously been added to the cultivation broth at 3%. The medium was filtered at pH 3 under 1 kg/cm^2 pressure to separate the cells. Table 21 shows results that demonstrate a similar tendency to the results obtained in flask tests. The highly-productivity strain 527–1 exhibited excellent characteristics, thus allowing the use of inexpensive glucose as a carbon source. The product concentration for this strain increases approximately 500 to 600 times higher than that of the wild strain with a low oxygen supply. The filtration efficiency was improved approximately twentyfold without by-products. These characteristics

Table 21. Properties of a strain improved from the biochemical engineering viewpoint[a].

Property	Wild strain	High-productivity strain (527–1)
Relative product concentration	1	500–600
Carbon source	Glycerol	Glucose
Oxygen supply ($k_L a$)	200 h^{-1}	50–100 h^{-1}
Filtration	20 liter/h/m^2	410 liter/h/m^2

[a] A 600-liter tank using commercially available natural organic substances.

suggest the suitability of the strain for industrial production by satisfying the biochemical engineering requirements.

3.3. Carbon Consumption and the Activity of TCA Cycle Enzymes in the High-Productivity Strain

The biosynthesis in secondary metabolism is affected by such factors as enzyme induction, enzyme activity, membrane transmittance, catabolites, final product, and the chemical and physical environments. Therefore, it is vitally important to make clear the law of cause and effect between those factors and the final product in order to succeed in the industrial production of microbial metabolites. Generally, the metabolism of microbial is economically performed for the growth and life of the cell. For large accumulation of metabolites, it is necessary to find the microorganisms with deficiency of metabolisms, and to purposefully regulate the metabolic cycles by altering the environment. A number of reports have discussed these matters from a viewpoint of metabolic regulation in breeding studies and actual productions (69–73).

Seto et al.[74] tried to add various radiolabeled compounds to cultures of a bialaphos-producing strain in order to make clear thier incorporation in AMPB (phosphenotricin) part of the bialaphos structure. They concluded that acetic acid (^{13}C-labeled) was incorporated at C-1 and C-2 and glucose (U-^{13}C-labeled) at C-3 and C-4, and the C-5 methyl (P-methyl) adjacent to P came from the methyl-radical of methionine. From those findings, the mechanism of bialaphos formation is presumed to be as follows: Glucose is converted to phosphoenol pyruvic acid (PEP), and it forms a C — P linkage. Next, the addition of acetic acid in the presence of acetyl-CoA and the addition of two alanines to the linkage may form bialaphos (Figure 29). Therefore, based on the assumption that there is a close relationship between glucose metabolism and bialaphos formation, carbon consumption and the activity of the TCA cycle enzymes in the high-productivity strain selected from the biochemical engineering viewpoint were examined comparing with those in the low-productivity strain.

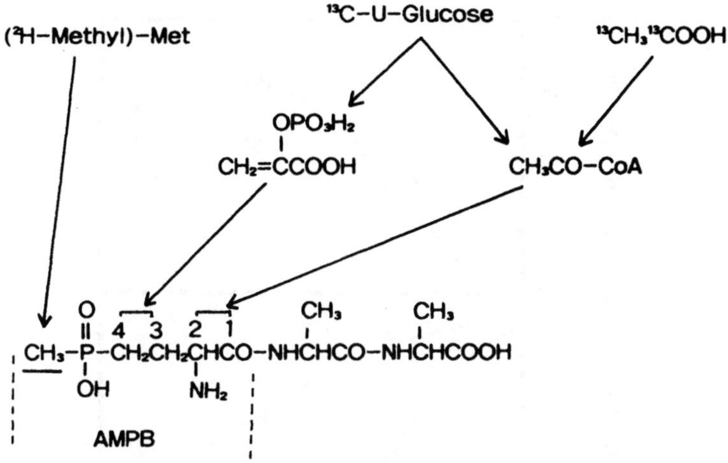

Figure 29. Addition of various radiolabeled compounds to the culture solution of a bialaphos-producing strain.

3.3.1. Sugar Consumption and Cell Weight

Figure 30 shows the time courses of the dry cell weight, sugar consumption, and bialaphos concentration of the low-productivity strain (24–15) in submerged cultivation using glycerol or glucose. Due to the effect of glucose catabolite regulation in the low productivity strain, bialaphos concentration in the glucose-based medium was only one-third of that in the glycerol-based medium. The cell weight and sugar consumption in the glucose-based medium were higher and faster than those in the glycerol-based medium. An increase in the bialaphos concentration occurs in both media when the growth is over. Figure 31 shows the comparison of the metabolisms of the low-productivity strain (24–15) and the high-productivity strain (527–1). In the glocose-based medium, the bialaphos concentration produced by the high-productivity strain, which has been released from glucose catabolite regulation, was five times higher than that of the low-productivity strain. The cell weight in the high-productivity strain was as low as two-thirds of that in the low-productivity strain. The growth phase of both strains ter-

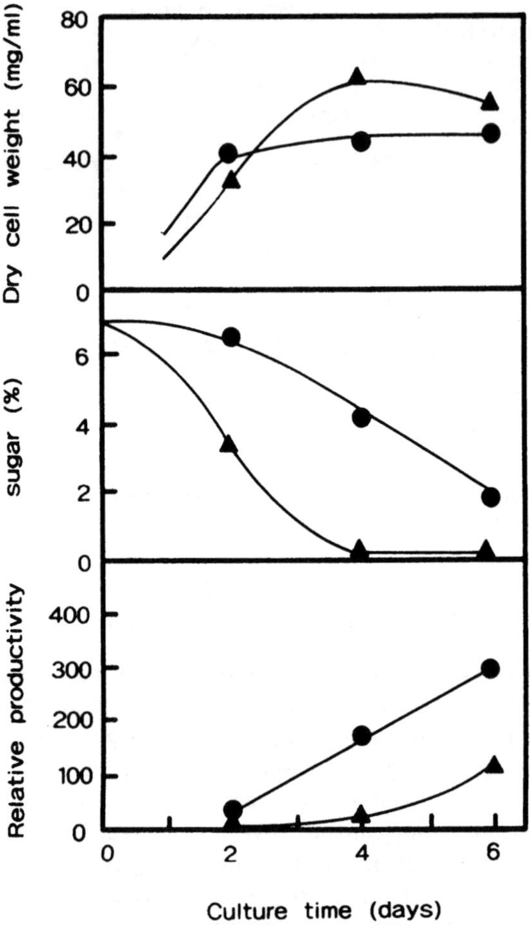

Figure 30. Time course of D.C.W., sugar, and productivity of bialaphos by a low-productivity strain in cultures using (●) glycerol and (▲) glucose.

minates at about the 48th hour. Also, consumption of sugar in the high-productivity strain was lower than that in the low-productivity strain. This means that the volume of cell mass is less and sugar consumption is slower when the higher producing strain or the

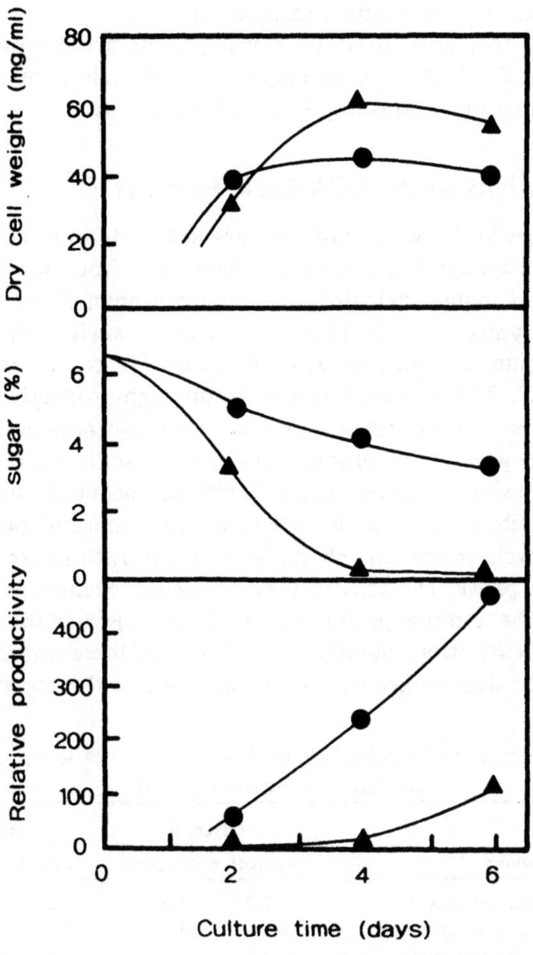

Figure 31. Time course of D.C.W., sugar, and productivity of bialaphos in cultures using a low-productivity strain (▲) and a high-productivity strain (●).

higher producing conditions are employed. These results suggest that regulation of the carbon metabolites in the high productive strain is stronger than that in the low-productivity strain, and, consequently, the stronger regulation might promote the flow of these compounds to the secondary metabolism.

In order to clarify these relationships, the activity of various enzymes in the TCA cycle and glyoxylic acid cycle were examined in relation to sugar consumption for both strains.

3.3.2. Activity of the TCA Cycle Enzymes

The changes in the activity of enzymes included in the TCA cycle, such as citrate synthase, aconitate hydratase, isocitrate dehydrogenase, and malate dehydrogenase, were examined for the low-productivity strain and the high-productivity strain in both glycerol-based medium and glucose-based medium. The results are shown in Table 22. The enzyme activities in the high-productivity strain were relatively lower than those in the low-productivity strain through the growth and production phases. Also, it was found that the activities of enzymes (citrate synthase, aconitate hydratase, isocitrate dehydrogenase) that are located at the initial position in the TCA cycle were relatively higher in the growth phase than the production phase. The activity of malate dehydrogenase, which was located at the terminal positions in the TCA cycle, had the opposite tendency to the prior. Namely, the activities of three enzymes other than malate dehydrogenase tend to be lower in the growth phase

Table 22. Tricarboxylic acid cycle enzyme activities[a] of low-productivity and high-productivity strains.

		Glycerol		Glucose	
Strain	Enzyme	Growth	production	Growth	production
Low	Citrate synthase	2.7	1.7	2.5	2.3
	Aconitate hydratase	398	298	672	373
	Isocitrate dehydrogenase	4.5	2.2	4.5	2.7
	Malate dehydrogenase	14.9	17.4	29.4	34.8
	(Relative productivity)	(314)		(100)	
High	Citrate synthase	-	-	1.1	0.8
	Aconitate hydratase	-	-	389	129
	Isocitrate dehydrogenase	-	-	2.5	1.2
	Malate dehydrogenase	-	-	12.4	14.9
	(Relative productivity)	(-)		(509)	

[a] Enzyme activity: μmole/min./mg protein.

than the production phase. Moreover, in the case of the low-productivity strain, the four enzyme activities in cells cultured in the glycerol-based medium were lower than those in cells cultured in the glucose-based medium.

3.3.3. Activity of Glyoxylic Acid Enzymes

In Table 23, the activities of glyoxylic acid cycle enzymes, such as isocitrate lyase and malate synthase, in both the low-productivity strain and the high-productivity strain cultured in the glucose-based medium were measured. The activities of the glyoxylic acid cycle enzymes tend to show the opposite relationship to those of the TCA cycle. Namely, the activities of both enzymes in the high-productivity strain were relatively higher than those in the low-productivity strain through the growth and production phases. The activities of both enzymes also increased in the production phase rather than in the growth phase.

Based on these findings, it was concluded that the high-productivity strain, compared with the low-productivity strain, made the consumption rate of sugar slower, cell mass less, the activities of the TCA cycle enzymes lower, and the activities of the glyoxylic acid cycle enzymes higher. These suggested that the flow to the TCA cycle from pyruvate or acetyl-CoA (substrate for bialaphos biosynthesis) was suppressed, and these substrates were efficiently directed to the secondary metabolism by the activation of the glyoxylic acid cycle in the highly productive strain. Thus,

Table 23. Glyoxylic acid cycle enzyme activities[a] of low-productivity and high-productivity strains.

Strain	Enzyme	Growth	Production
Low	Isocitrate lyase	3.3	3.9
	Malate synthase	1990	2080
	(Relative productivity)	(100)	
High	Isocitrate lyase	5.4	5.9
	Malate synthase	2280	4480
	(Relative productivity)	(509)	

[a] Enzyme activity: μmole/min./mg protein.

the high-productivity strain resulted in increasing bialaphos formation.

The scheme of bialaphos biosynthesis by *Steptomyces hygroscopicus* is illustrated in Figure 32. The biosynthesis of bialaphos involves processes of the conversion of glucose to a C—P linkage via phosphoenol pyruvic acid (PEP), addition of acetic acid in the presence of acetyl-CoA, and the addition of two alanines. Compared with the low-productivity strain, the high-productivity strain

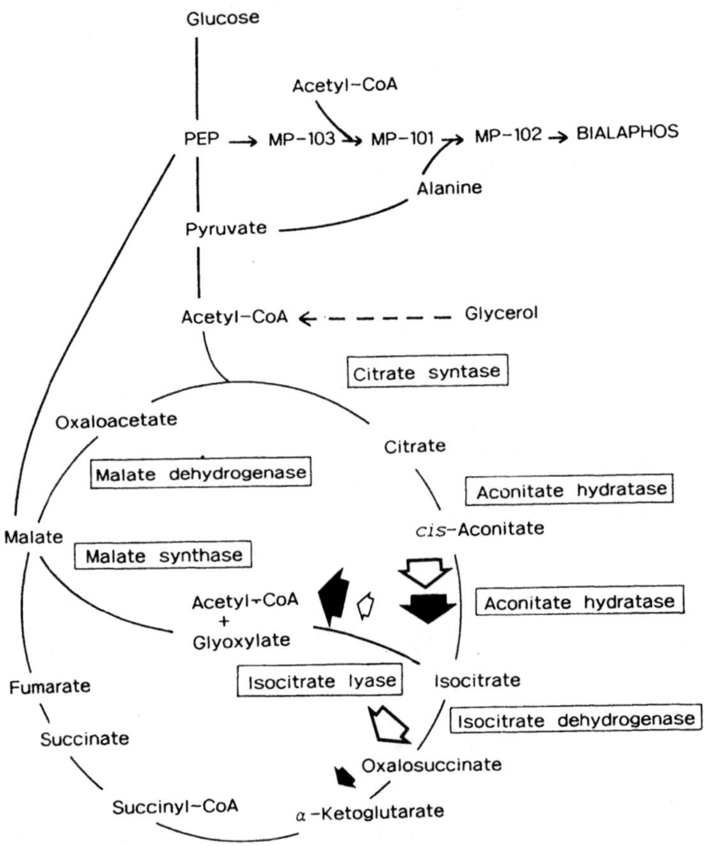

Figure 32. Scheme of bialaphos biosynthesis by *Streptomyces hygroscopicus*. ◊ : low-productivity strain; ◆ : high-productivity strain.

efficiently produces more bialaphos than cell mass from pyruvate or acetyl-CoA.

Hostalek[75] examined the activities of the TCA cycle enzymes in both the low-productivity strain and the high-productivity strain of *Streptomyces aureofciens* producing chlortetracycline. It was reported that the enzyme activities in the highly productive strain are relatively lower than that in the low-productivity strain through both the growth and production phases. Therefore, acetyl-CoA, a substrate for chlortetracycline, was mainly consumed for biosynthesis of the product so that the high-productivity strain had high productivity of chlortetracycline. This conclusion is in good agreement with our findings obtained in studies of bialaphos synthesis.

3.4. Effect of Oxygen Pressure on Bialaphos Synthesis in the High-Productivity Strain

It is well known that the oxygen supply excerts a considerable influence on production of aerobic fermentation such as antibiotics and amino acids. For example, in the glutamate fermentation[76], under a condition of sufficient oxygen supply, the pathways of bacterial metabolism are directed to the synthesis of a high yield of glutamate, whereas under a low oxygen supply, the metabolic pathways are divert from glutamate synthesis and produce more succinate or lactate, resulting in a low yield of glutamate. In cephalosporin fermentation,[77,78] production of penicillin N decreased and production of cephalosporin increased as the oxygen supply increased. As a further example, in tetracycline fermentation,[79] temporary interruption of the oxygen supply, even for only a short period, markedly inhibits the production. In this way, the oxygen supply is an important factor in the metabolism of aerobic organisms.

In Figure 27, we mentained that the high bialaphos-producing strain of *Streptomyces hygroscopicus* produced high yields of bialaphos under a condition of low oxygen supply, whereas an increased oxygen supply caused the production of bialaphos to decrease. In the bialaphos fermentation by the high-productivity strain, the reason for the high yield of the production under a low oxygen supply may be that a high oxygen supply is not required because the cell population per unit liquid culture volume is very small or because the respiratory activity per unit cell population is very low. On the other hand, the reason for the low yield of bialaphos under the condition of high oxygen supply may be that substrates required for the biosynthesis of bialaphos are consumed more for cell growth or that the high oxygen partial pressure causes suppression of enzymes related to secondary metabolism for bialaphos biosynthesis or causes accumulation of inhibitory substances.

This section investigates the biosynthesis of bialaphos by a high-productivity strain selected from the biochemical engineering viewpoint, by examining the effects of the oxygen partial pressure on the sugar consumption, activity of the TCA enzymes, and bialaphos production. Comparison is made with those in the low-productivity strain.

3.4.1. Oxygen Partial Pressure and Bialaphos Production

The high-productivity strain was cultured under various oxygen partial pressures by introducing compressed air or a mixed gas of air and nitrogen or oxygen using a new shaking apparatus of closed vessels that were pressure-proof and gas-suppliable, as shown in Figure 33. As shown in Table 24, elevation of the oxygen partial pressure to 0.42 atm by either compressed air or a mixed gas of air and oxygen decresed the production of bialaphos to 30–40% of control, ordinary culture (under an oxygen partial pressure of 0.21 atm). In contrast, supplying a mixed gas of air and nitrogen at 0.105 atm of oxygen partial pressure increased the production of bialaphos by 40%.

The effects of the starting substrates will be described later: When glucose was employed as the substrate, the suppressive effect of the high oxygen partial pressure on bialaphos production was observed only with the high-productivity strain, not with the low-productivity strain.

In regard to the method for regulating oxygen partial pressure, elevation by supplying either compressed air or air–oxygen mixed gas gave a similar bialaphos concentration provided that the oxy-

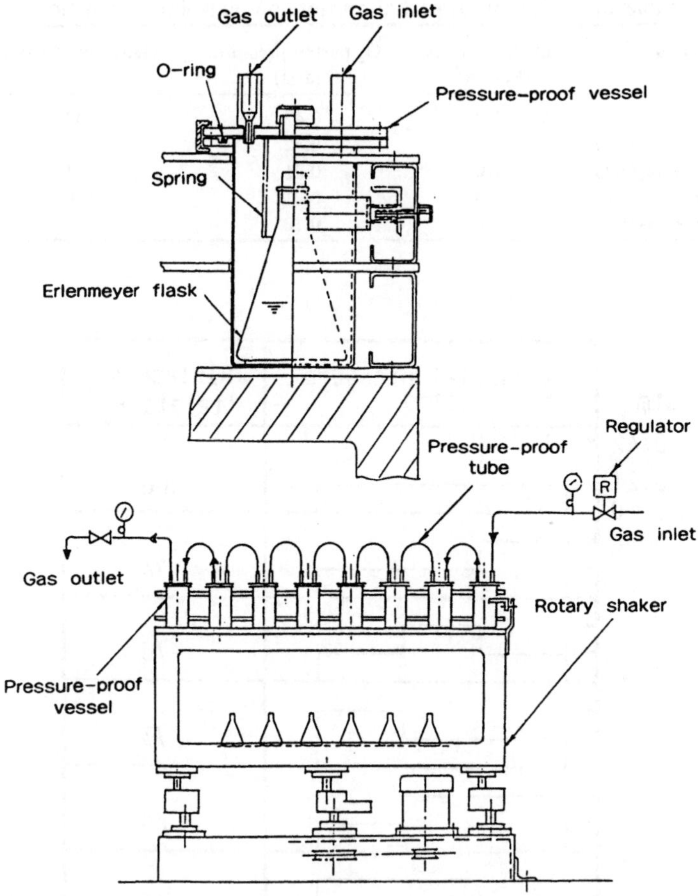

Figure 33. Equipment for increasing oxygen partial pressure in shake-culture.

gen partial pressure was identical (Table 24). Accordingly, in later experiments, elevation of the oxygen partial pressure was achieved using compressed air, which is the easiest technique.

Table 24. Effect of O_2 partial pressure on bialaphos production.

Inlet gas	Inner pressure (kg/cm^2)	O_2 partial pressure (atm)	Relative productivity
Air	1.0	0.42	53
Air + O_2(3:1)	0	0.42	58
Air + N_2(1:1)	1.0	0.21	111
Air	0	0.21	100
Air + N_2(1:1)	0	0.105	139

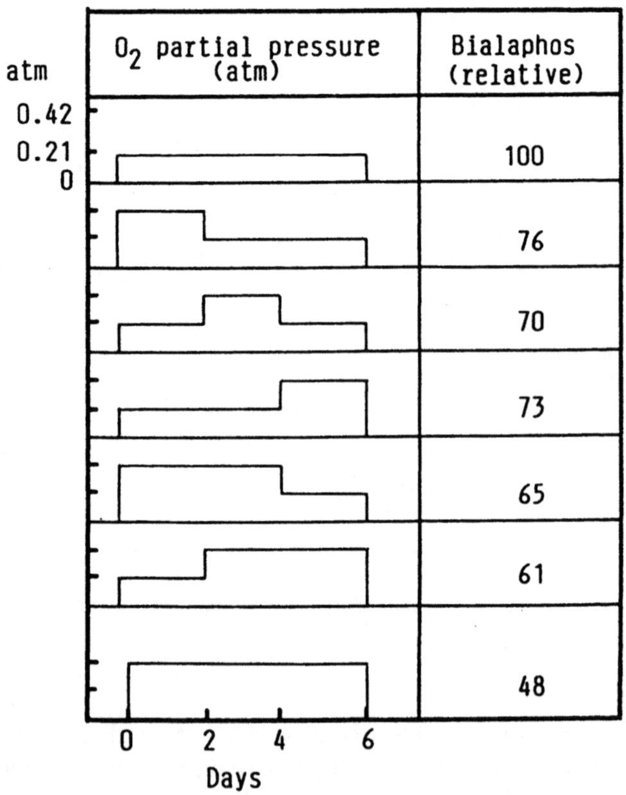

Figure 34. Effect of changes in oxygen partial pressure on bialaphos production.

3.4.2. Effect of Change in Oxygen Partial Pressure at Various Stages of Culture

As mentioned before, the oxygen partial pressure was a factor that affects the production of bialaphos. However, it was not yet clear which phase of the culture had been effected most prominently. Thus, the oxygen partial pressure was varied during three phases (growth phase: 0–2 days; early production phase: 2–4 days; and late production phase: 4–6 days), and the effect on bialaphos production was examined, as shown in Figure 34. Elevation of the oxygen partial pressure during any of those phases suppressed the production of bialaphos, and the longer the period of elevated oxygen partial pressure, the lower was the production of bialaphos. These findings demonstrated that keeping the oxygen partial pressure low throughout the whole period of bialaphos fermentation favors high bialaphos production.

3.4.3. Sugar Consumption, Cell Weight, and Respiratory Activity Under Elevated Oxygen Partial Pressure

Bialaphos production was markedly suppressed by elevating oxygen partial pressure in the culture flasks during either the growth or production phase. To clarify the mechanism of the suppression under a high oxygen partial pressure, variations along time of sugar consumption, cell weight, and respiratory activity were compared in the high-productivity and the low-productivity strains. Figure 35 shows the time courses of the dry cell weight, carbon source, and bialaphos yield under two different oxygen partial pressure conditions using the high-productivity strain. Although sugar consumption and dry cell weight of the low-productivity strain were not affected by the oxygen partial pressure (data not shown), it is seen from the figure that the sugar consumption was increased and the dry cell weight was somewhat increased under high oxygen partial pressure condition in the high-productivity strain.

The results suggest that an elevated oxygen partial pressure works to trigger the regulation of sugar metabolism so that the secondary metabolism was affected. To elucidate the lowering of bialaphos production under an elevated oxygen pressure, the respiratory acti-

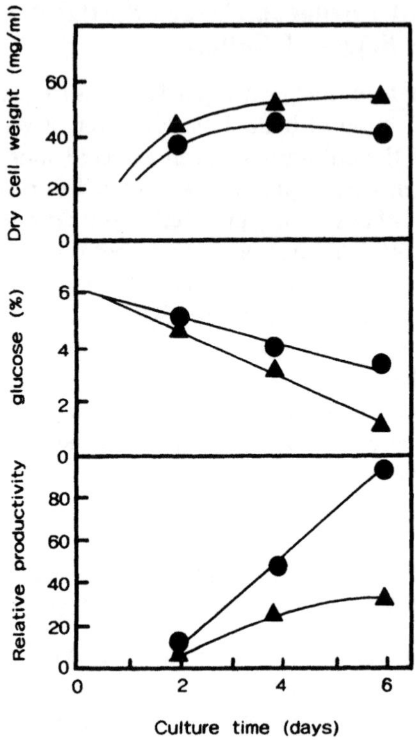

Figure 35. Effects of oxygen partial pressure on dry-cell weight, glucose, and bialaphos production. ●: 0.21 atm; ▲: 0.42 atm.

vity in cells cultured under high and ordinary oxygen partial pressures was studied and compared. The results are shown in Table 25. As is clear from the table, the respiratory activity (an index of cellular metabolic activity) in the high-productivity strain was higher than that in the low-productivity strain throughout the culture period.

The respiratory activity in the low-productivity strain decreased markedly as the culture proceeded, whereas in the high-productivity strain, the magnitude of the decrease was rather slight. When the oxygen partial pressure was elevated, the respiratory activity of the

Table 25. QO_2 of mycela harvested from cultures with increased oxygen partial pressure.

		Oxygen uptake rate (QO_2) ($\mu M\ O_2/mg \cdot h$)			
Strain		Low		High	
Age of mycelia		72 h	114 h	72 h	114 h
Oxygen partial pressure	0.21 atm	2.94	0.43	4.30	2.49
	0.42 atm	3.01	0.45	2.51	1.52

low-productivity strain did not change, whereas that of the high-productivity strain decreased greatly. These results suggest that the production of bialaphos and the respiratory activity (an index of aerobic metabolic activity) in the high-productivity strain are closely correlated, and the synthesis of bialaphos is apparently increased as the respiratory activity is augmented or as high respiratory activity is maintained. Therefore, this implies that, by expansion of the aerobic metabolic cycle, the amount of ATP required for alanylation and methylation (both essential steps in bialaphos biosynthesis) is increased. On the other hand, elevation of the oxygen partial pressure should work to lower the activity of the aerobic metabolic cycle in the high-productivity strain, thereby decreasing the amount of ATP required for the bialaphos biosynthesis.

3.4.4. Activity of TCA Cycle Enzymes Under a High Oxygen Partial Pressure

As mentioned before, it was found that the highly productive strain had decreased respiratory activity, accelerated sugar consumption, and increased cell mass under a high oxygen partial pressure. These facts led us to speculate that under the high oxygen partial pressure, the activities of enzymes of the primary metabolism change so that supply of substrates to the secondary metabolism becomes low.

To test the validity of this speculation, we investigated the rela-

Table 26. Effect of oxygen partial pressure on TCA cycle enzyme activities[a] of low-productivity and high-productivity strains.

		Oxygen partial pressure			
		0.21 atm		0.42 atm	
Strain	Enzyme	Growth	Production	Growth	Production
Low	Citrate synthase	2.5	2.3	2.5	2.1
	Aconitate hydratase	672	373	647	348
	Isocitrate dehydrogenase	8.2	2.7	9.9	7.2
	Malate dehydrogenase	24.9	34.8	22.4	35.9
	(Relative productivity)	(100)		(83)	
High	Citrate synthase	1.1	0.8	0.9	0.7
	Aconitate hydratase	398	129	448	192
	Isocitrate dehydrogenase	2.5	1.2	3.2	1.6
	Malate dehydrogenase	12.4	14.9	24.9	49.8
	(Relative productivity)	(509)		(183)	

[a] Enzyme activity: μmole/min./mg protein.

tionship between the activities of TCA cycle enzymes closely involved in sugar metabolism and the oxygen supply. Table 26 shows the changes in activity of four enzymes of the TCA cycle, that is, citrate synthase, aconitate hydratase, isocitrate dehydrogenase, and malate dehydrogenase, in both the low-productivity strain and high-productivity strain under two conditions of oxygen supply. The activity of these enzymes in the low-productivity strain was not changed at all by the change in the oxygen partial pressure through the growth and production phases. In the case of the high-productivity strain, the activities of all enzymes except citrate synthase under a high oxygen partial pressure (0.42 atm) were relatively higher than those under the ordinary partial pressure (0.21 atm) during both phases. Namely, the primary metabolism in the high-productivity strain was apparently influenced by an elevated oxygen partial pressure.

3.4.5. Activity of Glyoxylic Acid Cycle Enzymes Under a High Oxygen Partial Pressure

The changes in activity of the glyoxylic acid cycle enzymes (isocitrate lyase and malate synthase) in both the low-productivity strain and

Table 27. Effect of oxygen partial pressure on glyoxylic acid cycle enzyme activities[a] of low-productivity and high-productivity strains.

		Oxygen partial pressure			
		0.21 atm		0.42 atm	
Strain	Enzyme	Growth	Production	Growth	Production
Low	Isocitrate lyase	3.3	3.8	1.2	1.2
	Malate synthase	1990	2080	1810	1650
	(Relative productivity)	(100)		(83)	
High	Isocitrate lyase	5.4	5.9	1.1	1.4
	Malate synthase	2280	4480	2050	1050
	(Relative productivity)	(509)		(183)	

[a] Enzyme activity: μmole/min./mg protein.

high-productivity strain were mesured under two conditions of oxygen supply. The results are shown in Table 27. The activities of the glyoxylic acid cycle enzymes tend to show the opposite relationship to those of the TCA cycle. The activities of glyoxylic acid cycle enzymes under a high oxygen partial pressure (0.42 atm) were relatively lower than those under the ordinary partial pressure (0.21 atm) during both phases. These phenomena were more prominent in the high-productivity strain than in the low-productivity strain.

Based on these results, it became clear that under a high oxygen partial pressure, the activities of the TCA cycle enzymes in cells increased, whereas the activities of the glyoxylic acid cycle enzymes in cells decreased.

The scheme of bialaphos biosynthesis by *Streptomyces hygroscopicus* is illustrated in Figure 36. The biosynthesis of bialaphos includes processes of the conversion of glucose to a C—P linkage via phosphoenol pyruvic acid (PEP), addition of acetic acid in the presence of acetyl-CoA, and addition of two alanines. In the high-productivity strain, high oxygen partial pressure promotes the activities of TCA cycle enzymes so that both acetyl-CoA and pyruvate, the substrates in the synthesis of bialaphos, are supplied more to the primary metabolism, resulting in low production of bialaphos by the high-productivity strain.

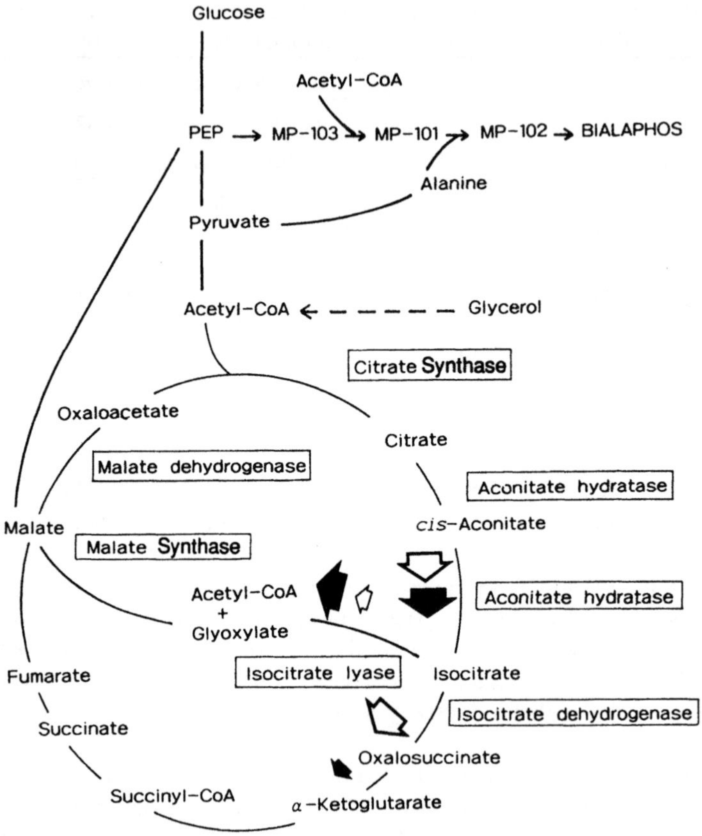

Figure 36. Scheme of bialaphos biosynthesis by *Strepomyces hygroscopicus*. ◊ : high oxygen partial pressure; ♦ : low oxygen partial pressure.

3.4.6. Bialaphos Production from Glycerol

The experiments described heretofore were done using glucose as the sole carbon source. We next investigated whether the suppression of bialaphos production by a high oxygen partial pressure is also seen with glycerol in place of glucose (Table 28). In the high-productivity strain, the suppression was observed in the glycerol

Table 28. Effect of oxygen partial pressure on relative bialaphos concentrations of the low-productivity (24.15) and high-productivity (527.1) strains P-1 medium with glucose or glycerol.

	Relative productivity			
Strain	Low		High	
O_2 partial pressure (atm)	0.21	0.42	0.21	0.42
Glucose	100	89	530	185
Glycerol	300	155	390	160

medium as well as in the glucose medium, but in the low-productivity strain, the suppression was observed only in glycerol medium. The low-productivity strain produced more bialaphos in glycerol medium than in glucose medium.

These findings suggest that the phenomenon of the suppression of bialaphos production by a high oxygen partial pressure is not dependent on the characteristics of strain; as production of bialaphos increased, the suppression by a high oxygen supply tended to become larger. Based on the facts described before, we conclude that high bialaphos production can be achieved under well-controlled oxygen supply conditions.

It is known that the maximum production of metabolites was obtained in acetoin and 2,3-butyleneglycol fermentation[80] by *Bacillus subtilis* and amino acid fermentation[81,82] such as phenylalanine, valine, and leucine when the activity of TCA cycle is inhibited in some degree by a limited oxygen supply. It is interesting to note that all of those metabolites are produced from pyruvate or phosphoenol pyruvate as starting substrates, as is bialaphos.

3.5. The Effect of High Oxygen Partial Pressure on Bialaphos Production Inhibition

As mentioned before, concentration of bialaphos by a high-productivity strain is greatly reduced under high oxygen partial pressure. As reasons for the concentration reduction under such a high oxygen partial pressure, the followings are speculated:

(a) production of bialaphos is reduced due to predominant shifting of substrate consumption from the bialaphos biosynthesis to cell growth, (b) enzymes for secondary metabolic biosynthesis are inhibited, and (c) inhibitory substances for production are accumulated. As to (a), we have confirmed that the speculation was reasonable.

This section deals with the culture solution obtained under high oxygen partial pressure, which was analyzed in order to clarify whether or not points (b) and (c) are involved.

3.5.1. Accumulation of Intermediates of Bialaphos Under Pressurized Culture

Concentration of bialaphos by the high-productivity strain is greatly reduced. To clarify the phenomenon of such a production inbibition under high oxygen partial pressure, the culture solution obtained under the conditions was analyzed. The high oxygen partial pressure was given by pressurized air through a compressor using a shaking apparatus described before.

Although bialaphos, AMPB, and two other unknown substances were present in the culture solution obtained under ordinary pressure (0.21 atm) and pressurized conditions, large amount of unknown substances that were likely to be biosynthetic intermediates of biapaphos were accumulated in cultures under pressurized condition (0.42 atm; Figure 37). From speculation on retention time, these unknown substances were anticipated to be biosynthetic intermediates MP-101 and MP-102, which are known to be accumulated in the culture solution without cobalt presence (Seto et al.[83]). In order to identify the unknown compounds, isolation and purification were carried out.

Compounds isolated were analyzed by ^{1}H-NMR to be identical to authentic MP-101 and MP-102, namely, one was demethyl bialaphos (MP-102) and the other dimethyl AMPB (dealaninated bialaphos) (MP-101). Because demethyl derivatives of bialaphos biosynthetic intermediates were accumulated, it was suggested that the methylation step was suppressed under high oxygen partial pressure.

To clarify the point, the following experiments were carried out.

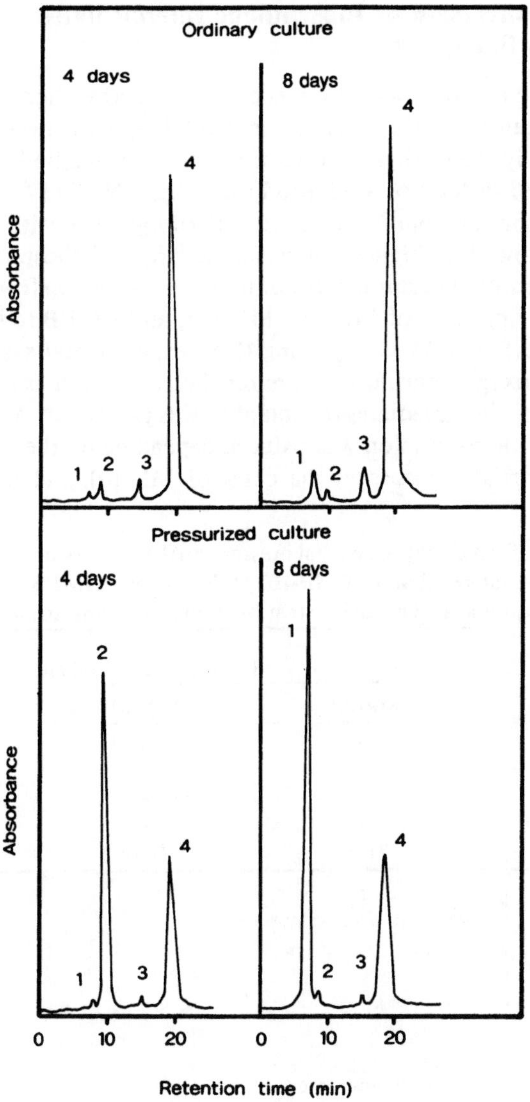

Figure 37. Amino acid analysis of the fermentation broth under ordinary culture (0.21 atm) and pressurized culture (0.42 atm). 1: MP-101; 21: MP-102; 3: AMPB; 4: bialaphos.

3.5.2. Conversion of Biosynthetic Intermediates to Bialaphos

In order to demonstrate suppression of the methylation step under high oxygen partial pressure, a methodology was investigated to trace clearly the methylation reaction only. The method was to use washed cell systems of a nonproducing strain NP-213[84,85] in which C—P formation ability is blocked. Microorgainism with 3 days of cell age showed the highest converting activity and phosphate buffer was preferable. The optimal conditions for conversion from biosynthetic intermediates of MP-101, MP-102, and AMPB to bialaphos were pH 6.5 and 32°C. By using these experimental systems, the effect of oxygen partial pressure on the conversion from various biosynthetic intermediates to bialaphos was examined. As shown in Table 29, the conversion was reduced depending on the increase of oxygen partial pressure in the cases of MP-101 and MP-102 as

Table 29. Effect of oxygen partial pressure on the conversion from various biosynthetic intermediates of MP-101, MP-102, and AMPB to bialaphos using a washed cell system by a nonproducing strain.

O_2 partial pressure (atm)	Biosynthetic intermediates		
	MP-101	MP-102	AMPB
0.1	130	135	101
0.21	100	100	100
0.315	78	76	98
0.42	43	48	102
0.84	Trace	Trace	94

Bialaphos:

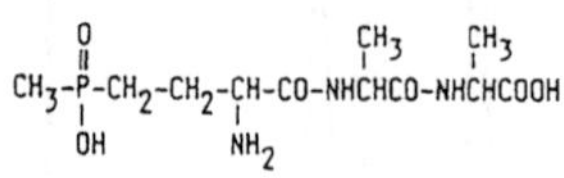

MP-101:

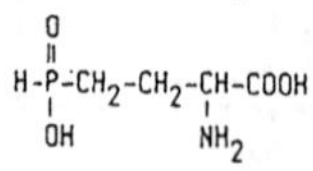

MP-102:

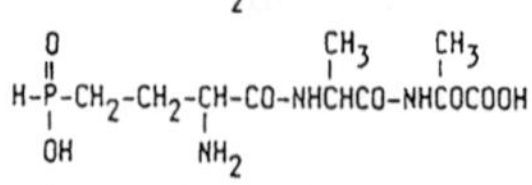

AMPB:

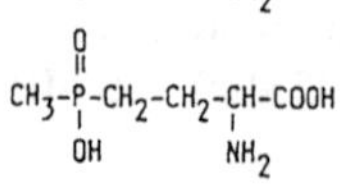

substrate, whereas proceeded in the case of AMPB. This indicates that amination and alanination in the biosynthesis of bialaphos are not affected by oxygen but methylation is.

In addition, the following causes that suppress methylation under high oxygen partial pressure would be assumed: when the partial pressure is high, (a) reduction of activities of methylation-related enzymes (coenzymes) due to oxidation, (b) decrease of production of the coenzymes during culture, (c) formation of inhibitors of methylation enzymes, and (d) suppression of methyl-donor supply are possible. However, it would be reasonable to assume that (a), (b), and (c) seem unlikely because bialaphos production started immediately after the oxygen partial pressure was reduced.

3.5.3. Addition of Various Methylation Agents

It was discussed before that the methyl-donor supply would be suppresed when the oxygen partial pressure is high. Hence, bialaphos productivity under high oxygen partial pressure was examined by the addition of methylation agents such as Vitamin B_{12}, methylcabalamine, methionine, and betaine into the culture solution. As shown in Table 30, biapaphos production recovered under high oxygen partial pressure (0.42 atm) in the cases of betaine and methionine. That is, bialaphos production was not reduced even under high partial pressure if the methyl donor is sufficiently sup-

Table 30. Addition of various methylation agents under high oxygen partial pressure.

	Oxygen partial pressure (atm)	
Methylation agent	0.21	0.42
None	100	62
[a]Vitamin B12	108	55
[a]Methylcobalamine	107	57
[b]Methionine	92	73
[b]Betaine	99	93
[c]$CoCl_2 \cdot 6H_2O$	102	57

[a] Vitamin B12, methylcobalamine: 0.005 mg/ml.
[b] Methionine, betaine: 1 mg/ml.
[c] $CoCl_2 \cdot 6H_2O$: 0.05 mg/ml.

plied. From this it was suggested that the supply of methyl donor is rate-limiting because bialaphos production is reduced under high oxygen partial pressure.

In bialaphos fermentation, it was demonstrated that oxygen partial pressure influences the primary and secondary metabolic processes; when oxygen partial pressure is increased, the TCA cycle is stimulated to change the substrate flow from bialaphos biosynthesis to cell-formation process and the supply of methyl donor is reduced to suppress the methylation reaction as the final step of bialaphos biosynthesis.

Based on the facts just described, we conclude that high bialaphos production can be achieved with a well-controlled oxygen supply. Scale-up for mass production bialaphos has been successfully achieved by using 100-kiloliter fermentor in which the effective culture method is employed to supply the minimal amount of oxygen for maintaining cell growth and activity.

CHAPTER 4

Novel Acyl Derivatives of Tylosin Produced by Microbial Transformation

Mitsuyasu Okabe and Rokuro Okamoto

4.1. Introduction

The family of macrolide antibiotics characterized by having a macrocyclic ring attached to aminosugars show potent antimicrobial activity against a wide range of Gram-positive bacteria. Furthermore, they are active to several kinds of Gram-negative bacteria and mycoplasma. Therefore, these antibiotics are widely used as pharmaceuticals for humans and animals (for domestic animals and fish culture).[86,87]

Since erythromycin was initially used in human chemotherapy in the United States in 1954, the following macrolide antibiotics have been introduced into clinical medicine: oleandomycin, leucomycin, spiramycin, josamycin, midecamycin, and their derivatives, such as triacetyloleandomycin, acetylspiramycin, and acetylmidecamycin. Due to such a rapid increase in number and quantity of macrolide antibiotics used for chemotherapy, macrolide-resistant strains have been clinically isolated at a high frequency as in the case of β-lactam antibiotics. Now, taking *Staphylococcus aureus* as an example, clinically isolated frequency of erythromycin-resistant strains remarkably increased from 1 to 45% between 1961 and 1981 in Japan.[88]

Furthermore, it is of considerable interest that *Staphylococcus pyogenes*, which show cross-resistance to all kinds of macrolides, have been increasing rapidly since 1970.[89,90] In such medical circumstances, urgent demand for novel macrolide antibiotics effective against macrolide-resistant strains thrust many researchers into the following lines of approach:

1. screening microorganisms to produce novel macrolide antibiotic components effective to macrolide-resistant strains
2. improving known antibiotics to new clinically valuable derivatives with chemical and/or microbiological modification.

Although numerous novel macrolide antibiotics and derivatives had been reported respectively, we could find few compounds that were effective to macrolide-resistant microorganisms. In this chapter, we review our screening system for 16-membered macrolide-transforming microorganisms, discovery of 3-acetyl-4″-isovaleryltylosin, which shows excellent antimicrobial activity against macrolide-resistant bacteria and mycoplasmas, its biological properties, its action mechanism, and its process development for commercial production.

4.2. Background

So far, more than 60 macrolide antibiotic components have been reported.[91] These antibiotics can be classified as 12-, 14-, or 16-membered ring macrolides according to the size of the macrocyclic lactone ring. Recently, 16-membered macrolides have been intensively studied. For example, Omura et al.[92] discussed the significant relationships between the chemical structures and the biological properties of leucomycins that were representatives of 16-membered macrolides. They concluded the following:

1. 3-hydroxyleucomycins are antimicrobially more active *in vitro*, but less satisfactory in serum level and toxicity, than their 3-acetyl counterparts.
2. The antibacterial and antimycoplasma activity increases with increasing size of the acyl group in the 4″-position, from acetyl, propionyl, butyryl, to isovaleryl.
3. Reduction of the 18-aldehyde group markedly decreases antimicrobial activity.

Furthermore, chemical and biologial transformations of macrolide compounds by such reactions as 3-acylation,[93] 9-acylation,[94,95] 14-hydroxylation,[96] 18-reduction,[96,97] 4″-deacylation and acylation,[96,100–102] 3″-hydroxylation,[103,104] reduction of the chromophore,[105, 106, 107] and removal and addition of the sugar moieties[108, 109] have been proved to result in more or less remarkable changes of the physicochemical and biological properties of macrolide antibiotics.

Depending on the absence or presence of methyl radicals at the 4 and 12 positions, the methyl group of 16-membered macrolide antibiotics is further classified into two subgroups[92]:

1. *Carbomycin–leucomycin subgroup.*
   This subgroup, which lacks methyl radicals, contains about 40 components, including leucomycin, spiramycin, YL-704, SF-837, espinomycin, maridomycin, and carbomycin. They are usually produced in a mixture of several analogues that share a common chromophore and differ in the position of acylation and the type of acyl group.
2. *Tylosin–chalcomycin subgroup.*
   This subgroup, which possesses methyl radicals, is composed of tylosin, anglomycin, cirramycin, rosamycin, chalcomycin, etc. In the tylosin–chalcomycin subgroup, comparatively few analogues are known, though the chances of acylation seem to be similar to or higher than those in the carbomycin–leucomycin subgroup. Accordingly, directed acylation of hydroxyl groups macrolide compounds of the tylosin–chalcomycin subgroup will provide new, and hopefully more useful, derivatives.

Because previous studies indicate the possibility of improving the physicochemical and pharmacological properties of macrolide antibiotics by various reactions, tylosin and anglomycin, which are useless for treatment of infections due to low blood levels in human, but show excellent *in vitro* antimicrobial activity against pathogenic bacteria and mycoplasmas, might be biochemically transformidable into clinically valuable macrolide.

Taking into consideration the previous discussions of the relationships between the chemical structures and the biological properties, we have established the screening system of macrolide-transforming microorganisms, with particular attention to acylation at the 3- and 4″-hydroxyl groups of 16-membered macrolides. Furthermore, we focused our attention on tylosin due to its excellent antimicrobial activity against macrolide-resistant strains: we have already obtained the data from a preliminary test that tylosin inhibits resistant strains at about twice the lower concentration than erythromycin, acetylspiramycin, josamycin, and midecamycin. We therefore chose tylosin as the starting material to prepare derivatives with improved activity.

4.3. Screening for 16-Membered Macrolide-Transforming Microorganisms

4.3.1. Screening System

Okamoto et al.[110,111] developed the following new screening system to find the microorganisms to transform known macrolides into new compounds effective against macrolide-resistant bacteria and mycoplasmas.

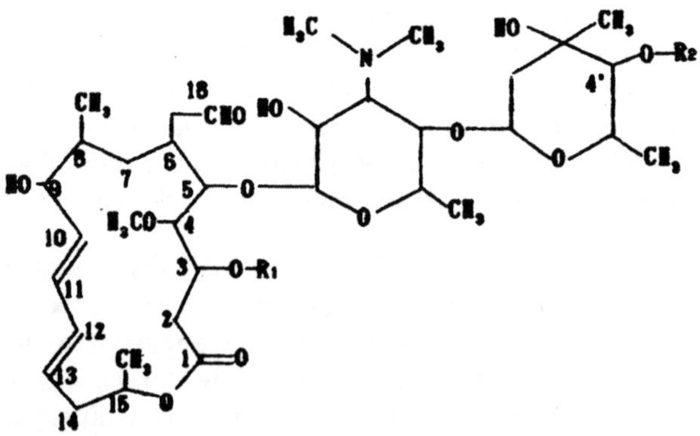

Substrate	R1	R2
Leucomycin A3	$COCH_3$	$COCH_2CH(CH_3)_2$
Leucomycin A1	H	$COCH_2CH(CH_3)_2$
Leicomycin U	$COCH_3$	H

Figure 38. Structures of substrates used for the screening of microbial transformation.

4.3.1.1. Substrate

As described previously, the final goal of our work was to find microorganisms to transform tylosin into clinically valuable new derivatives. Although tylosin might be used as starting substrate, leucomycin-group macrolides were used for primary screening because we had already established the TLC analytical system of those. Thus, three leucomycin analogues differing in the position and extent of acylation were used; they were leucomycin A_1 (3-hydroxy-4″-isovalerylleucomycin), leucomycin U (3-acetyl-4″-hydroxyleucomycin), and leucomycin A_3 (3-acetyl-4″-isovaleryl-leucomycin) (Figure 38).

4.3.1.2. Microorganisms

In total, 941 strains were screened, including 149 strains of bacteria, 275 strains of actinomycetes, and 517 strains of fungi. Most of them were type culture strains stocked in Mercian central research laboratories.

4.3.1.3. Cultivation and Reaction Conditions

Table 31 shows the cultivation and reaction system for the screening of macrolide-transforming microorganisms.

4.3.1.4. Detection of Reaction Product

Reaction products were qualitatively detected by thin-layer chromatography (TLC). The solvent systems for the development are shown in Table 32 and the TLC plates were air dried, dipped in 10% sulfuric acid, heated at 105°C for 15 min., and then examined for starting materials and transformation products. R_f values of leucomycin A_1, A_3, and U in different solvent systems are shown in Table 32.

After this quick preliminary check, the positive reaction mixture was analyzed quantitatively. A suitable volume of the ethyl acetate extract was spotted on the TLC plate together with known amounts of the corresponding substrate, and the plate was developed and colored as described before. The intensity of coloration of macrolide spots on the TLC plate was measured at 480 nm with a Shimazu

Table 31. Culture and reaction conditions for the screening of macrolide-transforming microorganisms.

Bacteria	
Slant culture	Nutrient agar, 36°C for 3–5 days.
Liquid culture	Medium A (peptone, 10 g; meat extract, 7 g; NaCl, 3 g; glucose, 7 g; tap water, 1000 ml; pH 7.0), shake cultured at 36°C for 24 h.
Reaction	Incubation overnight with shaking at 36°C after addition of substrate.
Actinomycetes	
Slant culture	ISP-2, 28°C for 5–10 days.
Liquid culture	Medium B (glucose, 20 g; soy bean meal, 5 g; yeast extract, 5 g; peptone, 5 g; K_2HPO_4, 1 g; $MgSO_4 \cdot 7H_2$, 0.05 g; tap water, 1000 ml; pH 7.0), shake-cultured at 28°C for 2–3 days.
Reaction	Incubated overnight with shaking at 28°C after addition of substrate.
Fungi	
Slant culture	ISP-2, 28°C for 5–10 days.
Liquid culture	Medium C (yeast extract, 5 g; malt extract, 20 g; K_2HPO_4, 0.5 g; $MgSO_4 \cdot 7H_2O$, 0.5 g; tap water, 1000 ml; pH 7.0), shake-cultured at 28°C for 2–3 days.
Reaction	Incubated overnight with shaking at 28°C after addition of substrate.

Table 32. Thin layer chromatography (TLC) system.

Thin-layer plate: Silicagel 60F254 (E. Merck, Darmstadt)

Solvent system	A (volume ratio)		B (volume ratio)
n-Hexane	30	Benzene	1
Acetone	10	Acetone	2
Methanol	8		
Benzene	25		
Ethyl acetate	20		

R_f values of the substrates in solvent systems A and B	A	B
Leucomycin U	0.12	0.24
Leucomycin A_1	0.33	0.61
Leucomycin A_3	0.48	0.70

Dual-Wavelength TLC-Scanner CS 900 and calibrated for the known concentration of the standard compounds.

4.3.2. Isolation and Identification of Transformed Compound

In order to isolate a substantial amount of transformation product, microorganisms were cultivated in jar fermentors containing 15 liters of the appropriate medium in Table 31. When the growth had reached maximum, the substrate was added to the culture broth. Aerobic incubation for microbial transformation was continued for a further 6 h under the same conditions as for the cultivation.

Transformation products were isolated by an established procedure for basic macrolide antibiotics. The detailed procedure for isolation and purification of transformation products are summarized in Figure 39.

Representative products detected in the primary screening were identified by gas chromatography, NMR and IR spectrometry, as well as TLC. The determination of constituent fatty acids of 16-membered macrolides by gas chromatography was made under the following conditions.

Gas chromatograph: Yanagimoto Model G-1800
Column: 1.5 m × 3 mm ϕ, packed with 10% PEAG and 1% H_3PO_4 on Chromosorb W
Injection temperature: 250°C
NMR spectra were measured in $CDCl_3$ with trimethyl silane as internal standards in a Hitachi R-24 high-resolution NMR spectrometer. For IR spectrometry, a Hitachi EPI-G2 infrared spectrometer was used.

4.4. Macrolide-Transforming Microorganisms and Characterization of the Types of Transformation

4.4.1. Results of the Primary Screening

Based on the TLC patterns of reaction product, microbial conversions are classified into five types, as shown in Table 33.

Reaction mixture (15 l)

↓ Filtered through a filter press at pH 4.

Filtrate

↓ Extracted with ethyl acetate at pH 8.0

Ethyl acetate extract (10 l)

↓ Back-extraction with 0.1M citrate buffer solution of pH 3.5 at 5°C

Aqueous extract (5 l)

↓ Re-extracted with ethyl acetate at pH 8 and 30°C

Ethyl acetete extract (2.5 l)

↓ Evaporated to dryness *in vacuo*

Pale yellow powder

↓

Silica gel column

↓ Slurried in benzene
Eluted stepwise with benzene: methanol (98:2 - 90:10)

Fraction of eluate

↓ Fractions containing the required product were identified by TLC and collected

Evaporated to dryness, dissolved in ethyl ether: isopropyl ether (4:1) and crystallized at 5°C or precipitated by addition of *n*-hexane.

Purified product

Figure 39. Procedure for the isolation and purification of transformation products.

Table 33. Types of transformation by microorganisms.

Type of transformation	Consumption of substrate[a]			Formation of product		Representative microorganism
				A[b]	B[b]	
I	Leucomycin A_3	−				
	Leucomycin A_1	+ +	+ +	0.48	0.70	*Streptomyces thermotolerans*
	Leucomycin U	+	+	0.48	0.70	ATCC 11416
II	Leucomycin A_3	+	+	0.33	0.61	*Bacillus subtilis*
	Leucomycin A_1	−				ATCC 14593
	Leucomycin U	−				
III	Leucomycin A_3	+ +	+ +	0.22	0.41	*Streptomyces piomogenes*
	Leucomycin A_1	±				A-428
	Leucomycin U	−				
IV	Leucomycin A_3	+ +	+	0.33	0.54	*Nocardia gardneri*
	Leucomycin A_1	+	±			
	Leucomycin U	−	±			IAM 105
V	Leucomycin A_3	+ +	+ +	0.12	0.24	*Bacillus megaterium*
	Leucomycin A_1	+	±			
	Leucomycin U	−				IAM 1032

[a] The amount of substrate consumed or product formed is shown by the following symbols: + +, ≧ 50 μg/ml; +, 10–50 μg/ml; −, < 10 μg/ml, and ±, several minor spots.
[b] For solvent systems A and B and R_f values of the substrates, see Table 32.

Type I is the transformation by genus *Streptomyces* of leucomycin A_1 and U, but not of leucomycin A_3. The R_f values with solvent systems A and B of the product from leucomycin A_1 and U coincided with those of leucomycin A_3. This reaction was clearly shown by *Streptomyces thermotolerans* ATCC 11416, *S. hygroscopicus* ATCC 21582, *S. narboensis var. josamyceticus* A-625, *S. fungicidicus subsp. espinomyceticus* A-656, *S. platensis* MCRL 0388, *S. kitasatoensis* NRRL 2486, and *S. mycarofaciens* SF-837. Interestingly, all the listed strains have been reported to produce macrolides having 3- and 4″-acyl groups such as carbomycin, maridomycin, and leucomycin. These observations suggest that reaction type I involves acylation of the 3- and 4″-hydroxyl groups.

Type II is the transformation by the genus *Bacillus* of leucomycin A_3, but not of leucomycin A_1 and U. The reaction product was identified as leucomycin A_1 by TLC with several solvent systems. Potent organisms are *Bacillus subtilis* IAM 1170, *B. subtilis* ATCC 14593, and *B. subtilis* B-14. As the involved reaction is thought to be deacylation of the 3-acyl group, the insusceptibility of leucomycin A_1 to the transformation merits further investigation.

Type III is the transformation by the genus of *Streptomyces* of leucomycins A_1 and A_3, but not of leucomycin U. The major product from leucomycin A_3 gave distinctly lower R_f values than the substrate on TLC with solvent systems A and B, indicating that the product was more hydrophilic. Transformation of leucomycin A_1 yielded several minor spots on TLC plates. As leucomycin U did not undergo this reaction, the possible direct or indirect involvement of the 4″-isovalery group was assumed. This type of transformation was effected by *Streptomyces piomogenes* A-428, *S. naganishi* IAM 0050, *S. antibiotics* ATCC 8663, *S. melanogenesis* IFO 12890, and *S. kasugaensis* A-401.

Type IV is the transformation of the three leucomycin substrates by the genera *Nocardia* and *Streptomyces*. Leucomycin A_3 yielded a more hydrophilic product, whereas leucomycin A_1 and U yielded several minor products that were hard to isolate because of chemical instability. Because all the substrates were susceptible, the aglycone of leucomycin might be involved. Included in this group are *Nocardia gardneri* IAM 105, *N. asteroides* A-158, *N. brasiliensis* A-159, *N. corollina* IFO 3338, *Streptomyces kanamyceticus* A-413, *S.*

almoquisti IFO 13105, *S. galilaceus* A-541, and *Actinomyces atrofaciens* ATCC 31104.

The final type (Type V) is a reaction common to lecomycin A_1 and A_3, but not leucomycin U. The identity of the R_f value of the product from leucomycin A_3 with that of leucomycin U confirmed the transformation to be 4″-deacylation. As most microorganisms have been shown to deacylate the 4″-acyl group of 16-membered macrolide antibiotics, the high frequency of detection of such strains in this screening was understandable. Under the culture and reaction conditions employed, strains of the genus *Bacillus* (*Bacillus megaterium* IAM 1032, *B. megaterium* IAM 1160, and *B. megaterium* IAM 1160, and *B. megaterium var. liquefaciens* B-296) were most active. As 4″-deacylase has been well characterized by several authors, enzymological study was not conducted.

4.4.2. Characterization of the Types of Transformation

For specification of the five types of transformation, the following representative strains and substrates were chosen, based on the primary screening:

Type	Microorganisms	Substrate
I	*Streptomyces thermotolerans* ATCC 11416	Leucomycin A_1 and U
II	*Bacillus subtilis* ATCC 14593	Leucomycin A_3
III	*Streptomyces piomogenes* A-428	Leucomycin A_3
IV	*Nocardia gardneri* IAM 105	Leucomycin A_3
V	*Bacillus megaterium* IAM 1032	Leucomycin A_3

A substantial amount of each transformation product was isolated from the culture broths of these microorganisms by the procedure described previously (Figure 32).

4.4.2.1. Type I Transformation

Products yielded from leucomycins A_1 and U by *Streptomyces thermotolerans* ATCC 11416 were identified as leucomycin A_3 by TLC, gas chromatography, IR spectrometry, and NMR spectrometry. Thus, type I transformation is concluded to be acylation of

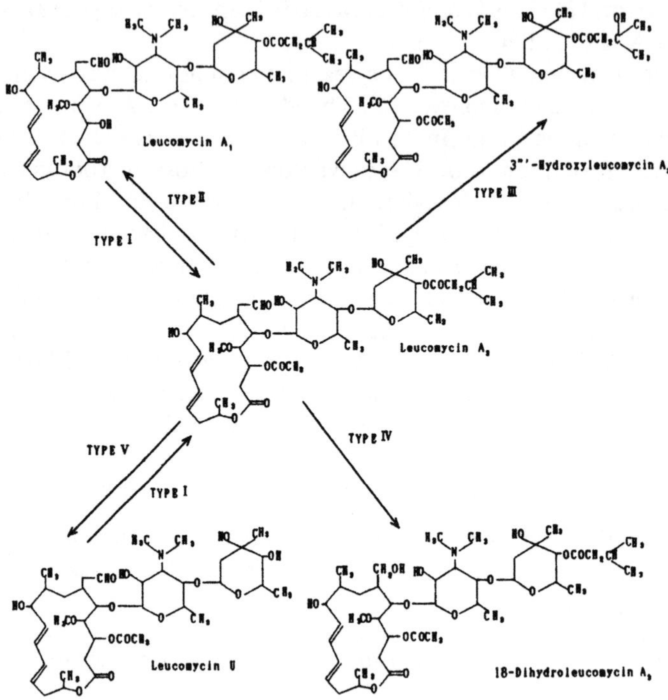

Figure 40. Structure relationships of the five types of microbial transformation in leucomycins.

the 3- and 4″-hydroxyl groups of leucomycin. Especially, *S. thermotolerans* ATCC 11416 acylated leucomycin A_1 at the 3-hydroxyl group and isovalerylated leucomycin U at the 4″-hydroxyl group, yielding A_3 in both cases (Figure 40).

4.4.2.2. Type II Transformation

Bacillus subtilis ATCC 14593 produced leucomycin A_1 from leucomycin A_3. The product was clearly identified as leucomycin A_1 by physicochemical analysis. Thus, type II transformation is concluded to be deacylation of the 3-acetyl group of leucomycin A_3 to give a 3-hydroxyleucomycin or leucomycin A_1 (Figure 40). This is a hitherto unknown type of the transformation.

4.4.2.3. Type III Transformation

Streptomyces piomogenes A-428 produced a more hydrophilic compound from leucomycin A_3. From the analysis of the NMR spectra, this product was considered to be 3‴-hydroxyleucomycin A_3 (a derivative of leucomycin A_3 hydroxylated at the β position of the isovaleryl group; Figure 40). The identity of the product with 3‴-hydroxyl leucomycin A_3 was confirmed by the TLC detection of acetic acid and β-hydroxyisovaleic acid in the hydrolysate. This type of transformation has been reported for maridomycin and leucomycin by Nakayama et al.[19]

4.4.2.4. Type IV Transformation

Nocardia gardneri IAM 105 converted leucomycin A_3 to a product with lower R_f value, which was identified as 18-dihydroleucomycin A_3. From the data of gas chromatogram and NMR spectra, this compound was also presumed to be 18-dihydroleucomycin A_3. 18-dihydroleucomycin A_3 was chemically synthesized by the reduction of leucomycin A_3 with $NaBH_4$ in ethyl alcohol. The product of the microbial transformation and chemically synthesized 18-dihydroleucomycin A_3 coincided in NMR spectrum, IR spectra, and melting point. This type of reductive transformation has long been known for macrolide compounds.[12,13]

4.4.2.5. Type V Transformation

As described before, 4″-deacylation, which was classified herein as type V transformation, has been reported in various species of microorganisms and in animals. As a representative strain, *Bacillus megaterium* IAM 1032 was employed for preparation of 4″-deacylleucomycin A_3 (Figure 40).

Table 34 summarizes the result of characterization of the five types of transformation. From the screening of microorganisms for transformation of macrolide antibiotics, we gained the general impression that among the organisms tested, actinomycetes were the most suitable to produce therapeutically interesting derivatives of 16-membered macrolides, as they far less frequently caused nonspecific decomposition of the substrate.

Table 34. Summary of microbial transformation.

Type of transformation	Substrate	Representative microorganisms	Product	Remarks
I	Leucomycin A_1	*Streptomyces thermotolerans* ATCC 11416	Leucomycin A_3	Acetylation of the 3-hydroxyl group
	Leucomycin U		Leucomycin A_3	Isovalerylation of the 4″-hydroxyl group
II	Leucomycin A_3	*Bacillus subtilis* ATCC 14593	Leucomycin A_1	Deacylation of the 3-acetyl group
III	Leucomycin A_3	*Streptomyces piomogenes* A-428	3‴-Hydroxyleucomycin A_3	Hydroxylation at the β position of the 4″-isovaleryl group
IV	Leucomycin A_3	*Nocardia gardneri* IAM 105	18-Dihydroleucomycin A_3	Reduction of the 18-aldehyde group
V	Leucomycin A_3	*Bacillus megaterium* IAM 1032	Leucomycin U	Deacylation of the 4″isovaleryl group

4.5. Discovery of New Acyl Derivatives of Tylosin

4.5.1. Preparation of Acyl Derivatives of Tylosin

In the course of screening for the microbiological transformation of 16-membered macrolides antibiotics, Okamoto et al.[25,26] found that *S. thermotorelans* ATCC 11416 could acylate the hydroxyl group at 3- and 4″-position of 16-membered macrolide antibiotics simultaneously at a high yield. In connection with the microbial acylatin of 16-membered macrolide antibiotics, *Streptomyces ambofaciens* NRRL 2420 had been reported to produce spiramycin II and III, having a 3-acetyl and 3-propionyl group, respectively, by acylation at the 3-hydroxyl group of spiramycin I.[93] Needless to say, this strain can only acylate at the 3 position of Spiramycin I,

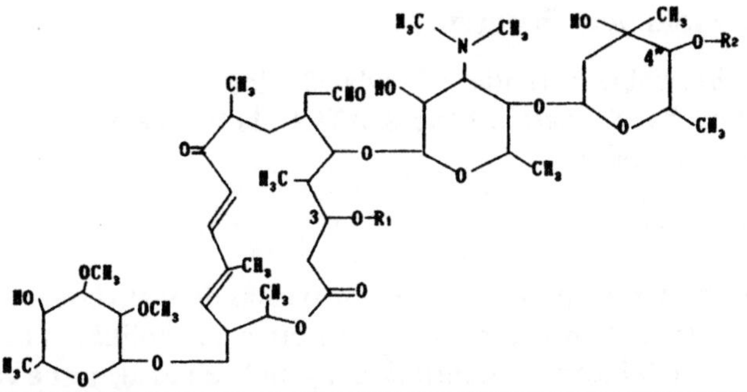

Compound		R_1	R_2
4"-Butyryltylosin	(2)	H	$COCH_2CH_2CH_3$
4"-Isovaleryltylosin	(3)	H	$COCH_2CH(CH_3)_2$
3-Acetyltylosin	(4)	$COCH_3$	H
3-Acetyl-4"-butyryltylosin	(5)	$COCH_3$	$COCH_2CH_2CH_3$
3-Acetyl-4"-isovaleryltylosin	(6)	$COCH_3$	$COCH_2CH(CH_3)_2$
3-Propionyltylosin	(7)	$COCH_2CH_3$	H
3-Propionyl-4"-butyryltylosin	(8)	$COCH_2CH_3$	$COCH_2CH_2CH_3$
3-Propionyl-4"-isovaleryltylosin	(9)	$COCH_2CH_3$	$COCH_2CH(CH_3)_2$
Tylosin(substrate)	(1)	H	H

Figure 41. Structures of new tylosin derivatives acylated by microbial conversion.

that is, one has a strict substrate specificity. On the contrary, *S. thermotolerans* ATCC 11416 can acylate a variety of 16-membered macrolides at 3 and/or 4″ position so that a wider variety of derivatives can be obtained by the proper combination of the substrate macrolide, the acyldonor such as propionyl CoA or isovaleryl CoA, and the reaction conditions. By the discovery of this strain, the next step of the screening program to find derivatives of tylosin effective against macrolide-resistant microorganisms was started. The following acyl derivatives of tylosin were prepared by microbial transformation by *S. thermotolerans* ATCC 11416[110,111]: 4″-butyltylosin, 4″-isovaleryltylosin, 3-acetyltylosin, 3-acetyl-4″-isovaleryltylosin, 3-acetyl-4″-butyryltylosin, 3-propionyl-4″-isovaleryltylosin, 3-propionyltylosin, 3-propyonyl-4″-isovaleryltylosin (Figure 41).

4.5.2. Biological Properties

The antimicrobial activities of acylated tylosins by microbial transformation by *S. thermotolelans* ATCC 11416 were examined *in vitro* and *in vivo*.[112]

4.5.2.1. MIC

The antibacterial spectra of the new tylosin derivatives are shown in Table 35 and 36 compparared with other macrolides. The MIC (minimum inhibitory concentration: μg/ml) values against a variety of microorganisms were measured by the agar dilution method, where BHI (brain–heart infusion) medium containing 10% horse blood was used. The strains listed in Table 35 are Gram-negative bacteria and macrolide-sensitive Gram-positive bacteria. The growth of the Gram-positive bacteria, but not of the Gram-negative bacteria, was inhibited by the eight acylated derivatives of tylosin as well as other macrolide antibiotics, that is, tylosin, erythromycin, and leucomycin A_3.

Tylosin and its acylated derivatives have similar antibacterial activity against macrolide-sensitive Gram-positive bacteria, as shown in Table 36. That is, introduction of an acyl group into the 3- and/or 4″ position(s) of tylosin did not increase the antibacterial activity against the macrolide-sensitive bacteria. In contrast, the macrolide-

Table 35. Antibacterial activity of new acylated tylosin derivatives with reference to known macrolides[a]

Test microorganisms	Compound[b]										
	(2)	(3)	(4)	(5)	(6)	(7)	(8)	(9)	(1)	EM	LM-A_3
Escherichia coli NIHJ	100	100	100	100	100	200	200	200	200	> 200	> 200
Pseudomonas fluorescens NIHJ B 254	> 200	> 200	> 200	> 200	> 200	> 200	> 200	> 200	> 200	> 200	> 200
Pseudomonas aeruginisa IAM 1133	> 200	> 200	> 200	> 200	> 200	> 200	> 200	> 200	> 200	> 200	> 200
Staphylococcus aureus Smith	0.39	0.39	0.39	0.39	0.39	0.39	0.39	0.39	0.39	0.19	0.39
Staphylococcus aureus FDA 209P	0.39	0.39	0.39	0.39	0.39	0.39	0.39	0.39	0.39	0.19	0.39
Bacillus cereus ATCC 9634	0.39	0.39	0.78	0.78	0.78	0.78	0.78	0.78	0.78	0.39	0.39
Bacillus subtilis ATCC 6633	0.39	0.39	0.39	0.78	0.39	0.39	0.78	0.39	0.39	0.39	0.39
Sarcina lutea ATCC 9341	0.19	0.19	0.19	0.19	0.19	0.19	0.19	0.19	0.19	0.19	0.19
Mycobacterium smegmatis ATCC 607	3.13	3.13	3.13	3.13	3.13	3.13	3.13	3.13	1.56	0.39	0.78
Streptococcus pyogenes NY-5	0.39	0.39	0.39	0.39	0.39	0.39	0.39	0.39	0.39	0.19	0.39

[a] MIC (μg/ml) was measured by the agar dilution method in BHI medium containing 10% horse blood.
[b] Compound: (2) 4″-butyryltylosin, (3) 4″-isovaleryltylosin, (4) 3-acetyltylosin, (5) 3-acetyl-4″-butyryltylosin, (6) 3-acetyl-4″-isovaleryltyltylosin, (7) 3-propionyltylosin, (8) 3-propionyl-4″-butyryltylosin, (9) 3-propionyl-4″-isovaleryltylosin, (1) tylosin, EM: erythromycin, LM-A_3: leucomycin A_3.

Table 36. Antibacterial activity of new acylated tylosin derivatives with reference to known macrolide antibiotics against macrolide-resistant bacteria[a].

	Compound[b]											
Test microorganisms	(2)	(3)	(4)	(5)	(6)	(7)	(8)	(9)	(1)	EM	LM-A_3	LM-U
Staphylococcus aureus[c1] MS 87100	50	50	> 200	50	25	> 200	50	25	> 200	> 200	> 200	> 200
Staphylococcus aureus[c2] MS 9610	100	100	> 200	100	100	> 200	200	100	> 200	> 200	> 200	> 200
Staphylococcus aureus[c3] MS 9931	12.5	12.5	100	25	12.5			25	200	> 200	> 200	
Staphylococcus aureus[c3] MS 9937	12.5	12.5	100	25	12.5			25	100	200	200	200
Streptococcus pyogenes[c4] MH 604	12.5	12.5	200	25	12.5			25	> 200	> 200	> 200	> 200
Streptococcus pyogenes[c4] MH 749	12.5	12.5	200	25	12.5			25	> 200	> 200	> 200	> 200
Streptococcus pyogenes[c4] MH 771	3.1	3.1	100	12.5	3.1	100	3.1	3.1	> 200	> 200	> 200	> 200

[a] MIC (μg/ml) was measured by the agar dilution method in BHI medium containing 10% horse blood.
[b] Compound: (2) 4″-butyryltylosin, (3) 4″-isovaleryltylosin, (4) 3-acetyltylosin, (5) 3-acetyl-4″-butyryltylosin, (6) 3-acetyl-4″-isovaleryltylosin, (7) 3-propionyltylosin, (8) 3-propionyl-4″-butyryltylosin, (9) 3-propionyl-4″-isovaleryltylosin, (1) tylosin, EM: erythromycin, LM-A_3: leucomycin A_3, LM-U: leucomycin U.
[c] Resistance pattern: [c1] PC-G·OTC·CBM-A, [c2] PC-G·CBM-A, [c3] OTC·CBM-A, [c4] PC-G·CEX·OTC·CBM-A (PCG: benzylpenicillin, CEX: cephalexin, OTC: oxytetracycline, CBM-A: carbomycin A).

resistant strains of *Staphylococcus* and *Streptococcus* displayed increased sensitivity to the new acylated tylosin derivatives and the 4″-acylation increased the antibacterial activity to a greater extent than the 3-acylation. For example, 4″-isovaleryltylosin shows stronger antimicrobial activity than 3-acetyl tylosin and tylosin. Thus, one may conclude that the 4″-acyl group plays a significant role in the antibacterial activity of the tylosin derivatives against macrolide-resistant bacteria in this series. However, in leucomycins, the 4″-acyl group seems to have no important effect on the antimicrobial activity with regard to resistant bacteria: leucomycin U is as inactive as leucomycin A_3. It is therefore probable that tylosin in itself has a latent specific activity against macrolide-resistant bacteria, which is expressed by 4″-acylation. From the viewpoint of antimicrobial activity to macrolide-resistant microorganisms, 3-acetyl-4″-isovaleryltylosin seemed to be most promising as a chemotherapy agent. The physicochemical properties of 3-acetyl-4″-isovaleryltylosin was reported elsewhere.[113]

4.5.2.2. Antimycoplasmal Activity

Tylosin is currently used for the treatment of mycoplasma infection in animals. The antimycoplasmal activity of 3-acetyl-4″-isovaleryltylosin and the reference antibiotics was examined against a sensitive strain (KP-13) and the macrolide-resistant strain (E-5, E-11, A-68, and A-72) of *Mycoplasma gallisepticum* (Table 37).

Table 37. Antimycoplasmal activity of 3-acetyl-4″-isovaleryltylosin in comparison with known antibiotics against Mycoplasma gallisepticum.

Compounds	MIC[a](μg/ml)				
	Mycoplasma gallisepticum				
	E-5	E-11	A-68	A-72	KP-13
3-acetyl-4″-isovaleryltylosin	0.62	0.31	0.31	0.31	0.02
Tylosin	10	5	2.5	2.5	0.02
Erythromycin	100	100	100	100	0.1
Spiramycin	100	100	100	100	0.1
Chloramphenicol	10	2.5	25	10	25
Oxytetracycline	0.62	0.31	1.3	0.62	0.62

[a] MIC was measured with the tube dilution method (10^5–10^6 CFU/ml).

Clearly from the table, the effect of 4″-acylation enhanced the antimycoplasmal activity more markedly than the antibacterial activity.

4.5.2.3. Stability in Artificial Gastric Juice

Because most macrolide antibiotics are often administered in the treatment of infections, the stability at a pH as low as 1.0 is important. The stability of 3-acetyl-4″-isovaleryltylosin in artificial gastric juice (ph 1.2) was examined, testing the activity by bioassay and TLC methods (Table 38). Tylosin was rapidly decomposed to give desmycosin, which was still bioactive. Leucomycin A_3 and 3-acetyl-4″-isovaleryltylosin on the other hand were slowly transformed to demycarosylleucomycin and 3-acetyldesmycosin, respectively. Thus, 3-acetyl-4″-isovaleryltylosin was shown to be more stable in the stomach than tylosin.

Table 38. Stability of 3-acetyl-4″-isovaleryltylosin in artificial gastric juice[a].

Compound	Incubation (h)	Amount of antibiotic remaining	
		TLC assay (%)	Bioassay (mm)
Tylosin	0	100	17.7
	2	t[b]	17.0
	4	t	17.0
	6	t	15.4
3-Acetyl-4″-isovaleryltylosin	0	100	16.5
	2	70	16.5
	4	38	15.9
	6	13	15.3
Leucomycin A_3	0	100	15.8
	2	58	15.6
	4	34	12.4
	6	13	8.1

[a] Drug was incubated at a concentration of 1000 μg/ml for the indicated periods of time at 37°C. The amount of the antibiotic remaining was measured by thin-layer chromatography (TLC) and by the disc assay with *Bacillus subtilis* ATCC 6633.
[b] t = trace.

4.5.2.4. Blood Levels of Tylosin and Acylated Tylosins after Oral Administration in Mice and Rabbits

Tylosin gives poor blood levels after oral administration. One purpose of its acylation is to obtain a derivative that can give a high blood level. After a single dose of tylosin, 3-acetyl or 3-acetyl-4″-isovaleryltylosin was given orally to mice and rabbits, the blood samples were collected at the indicated times, and the concentrations of the antibiotics were bioassayed with *Sarcina lutena* ATCC 9341 (Table 39). In mice, 3-acetyl-4″-isovaleryltylosin provided a markedly higher level than tylosin and 3-acetyltylosin. This was more clearly confirmed in rabbits. Because 3-acetyltylosin was as poor as tylosin, the improved blood level with 3-acetyl-4″-isovaleryltylosin could be ascribed to 4″-isovalerylation.

4.5.2.5. Treatment of Experimental Infections with *Staphylococcus aureus* Smith

Therapeutic effect of orally given 3-acetyl-4″-isovaleryltylosin was examined on mice systemically infected with *Staphylococcus aureus* Smith (Table 40). It is apparent that 3-acetyl-4″-isovaleryltylosin is superior to tylosin in the *in vitro* effect. This agrees with the improved blood levels. Multiple doses of 3-acetyl-4″-isovaleryltylosin showed a good effect in preventing infection.

4.5.3. Action Mechanism of 3-Acetyl-4″-isovaleryltylosin

The mechanism of the effect of 3-acetyl-4″-isovaleryltylosin against macrolide-resistant strains was intensively studied by Tsuchiya et al.[114,115] They examined whether this new derivative is taken up into resistant cells more efficiently than parent compound and whether this one inhibits protein synthesis on ribosomes of resistant strains. *Staphylococcus aureus* MS-9610 and MS-8710 strains were used throughout this study, because the former showed the highest resistance against all macrolide antibiotics and the latter showed the widest range of resistance covering drugs other than macrolides.

Table 39. Blood levels of 3-acetyl-4″-isovaleryltylosin after oral administration into mice and rabbits[a].

	Compounds	Dose (mg/kg)	Blood levels (μg/ml) 15 min.	30 min.	1 h	2 h	3 h
Mouse	Tylosin	100		t	0.45	t	
	3-Acetyltylosin	100		t	0.5	t	
	3-Acetyl-4″-isovaleryltylosin	100		9.5	14.0	2.5	1.8
		200	14.7	23.2	13.7	7.5	
Rabbit	Tylosin	50	t				
		200		0.4	t		
	3-Acetyl-4″-isovaleryltylosin	50		5.2	1.9	1.0	t
		200		21.3	18.5	6.2	3.6

[a] Concentration of the drug in the blood samples was disc-assayed with *Sarcina lutea* ATCC 9341.
[b] t = trace.

Table 40. Therapeutic effects of orally given 3-acetyl-4″-isovaleryltylosin in mice infected systemically with *Staphylococcus aureus* Smith.

Compound		Dose (mg/kg)	Survived/Treated				
			Day 1	Day 2	Day 3	Day 4	Day 5
		400	6/10				6/10
	Tylosin	300	6/10	5/10			5/10
		200	4/10	3/10			3/10
		100	3/10	1/10			1/10
Experiment 1[a]	3-Acetyl-4″-isovaleryltylosin	400					10/10
		300	7/10				7/10
		200	4/10	3/10			3/10
		200	1/10				1/10
	Untreated		0/10				0/10
		400 × 4					5/5
	Tylosin	100 × 4	1/5	0/5			0/5
		25 × 4	0/5				0/5
Experiment 2[b]	3-Acetyl-4″-isovaleryltylosin	400 × 4					5/5
		100 × 4					5/5
		25 × 4		3/5	2/5		2/5
	Untreated			0/5			0/5

[a] Experiment 1: Challenge dose 50 × LD_{50} in 5% mucin, i.p.; treatment 1-h infection.
[b] Experiment 2: Challenge dose 20 × LD_{50} in 5% mucin, i.p.; treatment 1, 4, 8 and 24 h after infection.

4.5.3.1. Effect of Cellular Macromolecule Synthesis

To measure the effect of 3-acetyl-4″-isovaleryltylosin on synthesis of cellular DNA, RNA, and protein, cells were labled with radioactive thymidine, uridine, and leucine, respectively, in the presence or absence of 3-acetyl-4″-isovaleryltylosin. As shown in Figure 42, the incorporation of leucine into a TCA-insoluble fraction of MS-8710 and MS-9610 was strongly inhibited by the addition of 3-acetyl-4″-isovaleryltylosin at 25 and 50 μg/ml, respectively. Tymidine and uridine incorporations were not inhibited (data not shown).

4.5.3.2. Uptake of 3-Acetyl-4″-isovaleryltylosin by Intact Cell

Uptake of 3-^{14}C-acetyl-4″-isovaleryltylosin into cells of the test microorganisms were measured by the membrane filter method.[114] The results shown in Figure 43, indicated, first of all, that decreased permeability was at least partly responsible for the resistance of *S. aureus* strains MS-8710 and MS-9610; the uptake of 3-^{14}C-acetyltylosin in resistant cells was significantly less than that of the sensitive strain(s). The degree of resistance of strain MS-9610 was stronger than strain MS-8710 and the uptake by the former

Table 41. Inhibitory activity of unlabeled 3-acetyl-4″-isovaleryltylosin into intact cells of *Staphylococcus aureus* in comparison with those of other compounds[a].

Concentration of unlabeled compounds (μg/ml)		Inhibition % compared with control		
		FDA 209P	MS-8710	MS-9610
Tylosin	20	41	40	35
	200	59	63	46
3-Acetyltylosin	20	39	49	29
	200	72	70	58
3-Acetyl-4″-isovaleryltylosin	20	66	68	54
	200	90	88	85

[a] Prior to the addition of 3-^{14}C-acetyltylosin to 0.4 ml of the cell suspension of *S. aureus*, 0.1 ml of aqueous solution of an unlabeled test compound was added to the cell suspension. After preincubation at 37°C for 10 min., 3.05 μg of 3-C^{14}-acetyltylosin (7.63 mCi/mmole) was added to the cell suspension and incubated for 20 mins. ^{14}C uptake of control (without unlabeled compounds) was 116 ng for FDA 209P, 82 ng for MS-8710, and 52 ng for MS-9610 per mg dry-weight cells, respectively. Inhibition was expressed in percent of control without inhibitor.

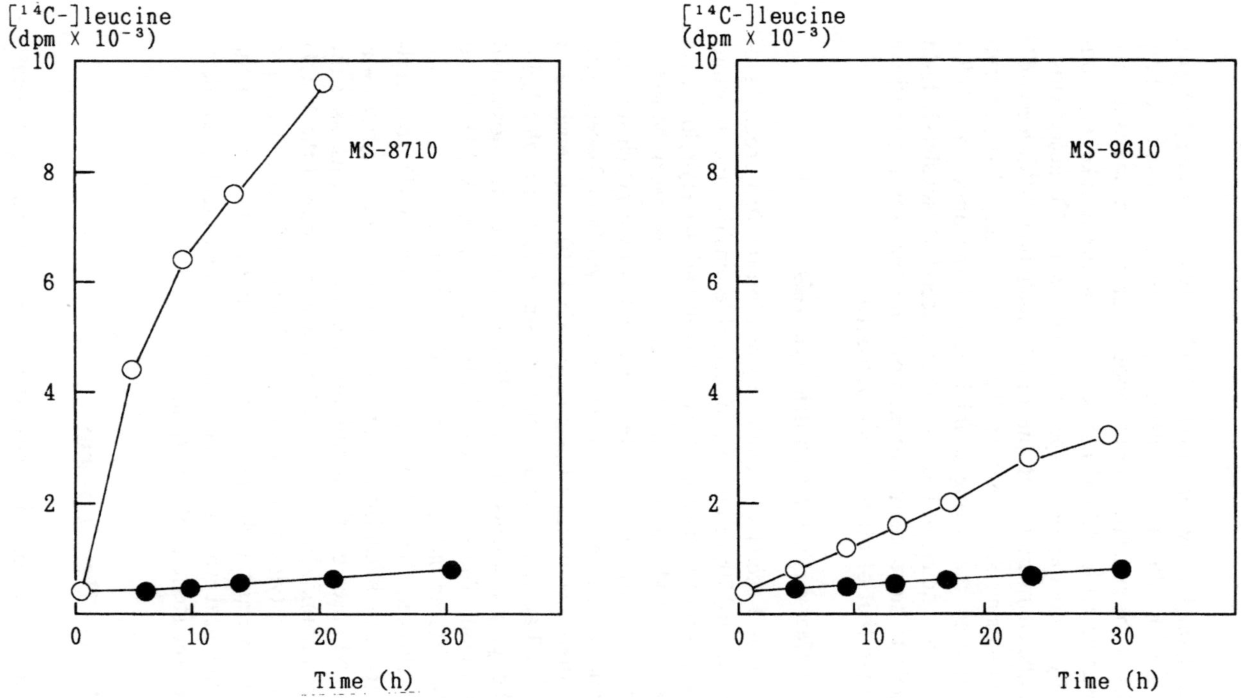

Figure 42. Inhibition of protein synthesis in intact cells of resistant strains MS-8710 and MS-9610. 3-acetyl-4″-isovaleryltylosin was added to give a final concentration of 25 μg/ml for MS-8710 and 50 μg/ml for MS-9610, respectively. ○: Control; ●: 3-acetyl-4″-isovaleryltylosin.

was less than that by the latter. 3-^{14}C-acetyl-4″-isovaleryltylosin, compared with 3-^{14}C-acetyltylosin, showed better uptake especially into cells of the resistant strains; about four times as much uptake was observed with MS-9610. Specificity of the uptake was confirmed by a competition experiment between 3-^{14}C-acetyltylosin and unlabeled members of the family including 3-^{14}C-acetyl-4″-isovaleryltylosin. As shown in Table 41, 3-acetyl-4″-isovaleryltylosin showed the strongest inhibition of the uptake of 3-^{14}C-acetyltylosin. The result indicates that 3-acetyl-4″-isovaleryltylosin can penetrate more easily into cells of both resistant and sensitive strains than that of other tylosin derivatives.

4.5.3.3. Effect on *In Vitro* Protein Synthesis

The effect of 3-acetyl-4″-isovaleryltylosin on *in vitro* protein synthesis on washed 70S-ribosomes of macroide-sensitive and -resistant strains was studied comparing wth tylosin and 3-acetyltylosin. As shown in Figure 44, the protein synthesis on ribosomes of MS-9610 cells was partially inhibited by 3-acetyl-4″-isovaleryltylosin, but little or no inhibition was demonstrated by tylosin or 3-acetyltylosin. On the other hand, ribosomes of MS-8710 were sensitive to tylosin and 3-acetyltylosin as well as to 3-acetyly-4″-isovaleryltylosin, although intact cells of MS-8710 were resistant to tylosin and 3-acetyltylosin. These results indicated that strains MS-9610 and MS-8710 differed in the mechanism of resistance.

In conclusion, the microbial activity of 3-acetyl-4″-isovaleryltylosin against clinical isolates of multidrug-resistant *S. aureus* was found to be based on the inhibition of protein synthesis, as is the parent compound. Introduction of a isovalerylgroup at the 4″-OH position of the tylosin increased the efficiency of penetration into both sensitive and resistant strains of cells by alternation in the physicochemical properties, that is, hydrophobicity and/or an alternation of the steric conformation.

4.6. Process Development

In the early stage of this study, the conversion yield (dry weight of 3-acetyl-4″-isovaleryltylosin produced/amount of tylosin added)

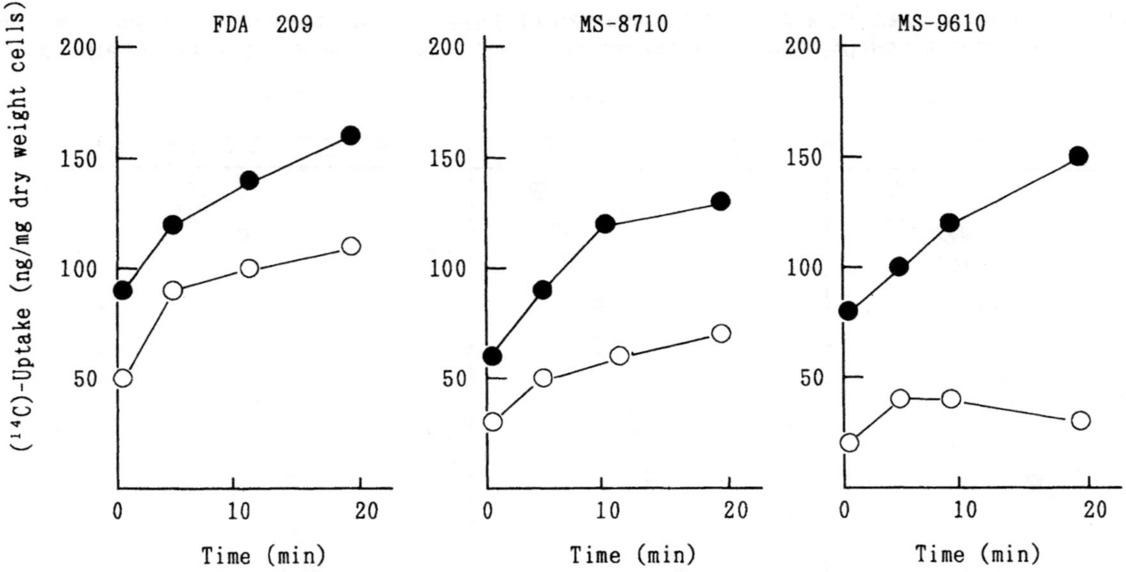

Figure 43. Uptake of ^{14}C-tylosin derivatives into intact cells of FDA 209P (a sensitive strain) and MS-8710, MS-9610 (resistant strains) of *Staphylococcus aureus*. Either 3.05 μg/ml of 3-^{14}C-acetylosin or 2.5 μg/ml of 3-^{14}C-acetyl-4″-isovaleryltylosin was added to 1.5 ml of the cell suspension. ●: 3-^{14}C-acetyl-4″-isovaleryltylosin (8.34 mCi/mmole); ○: 3-^{14}C-acetylosin (7.63 mCi/mmole).

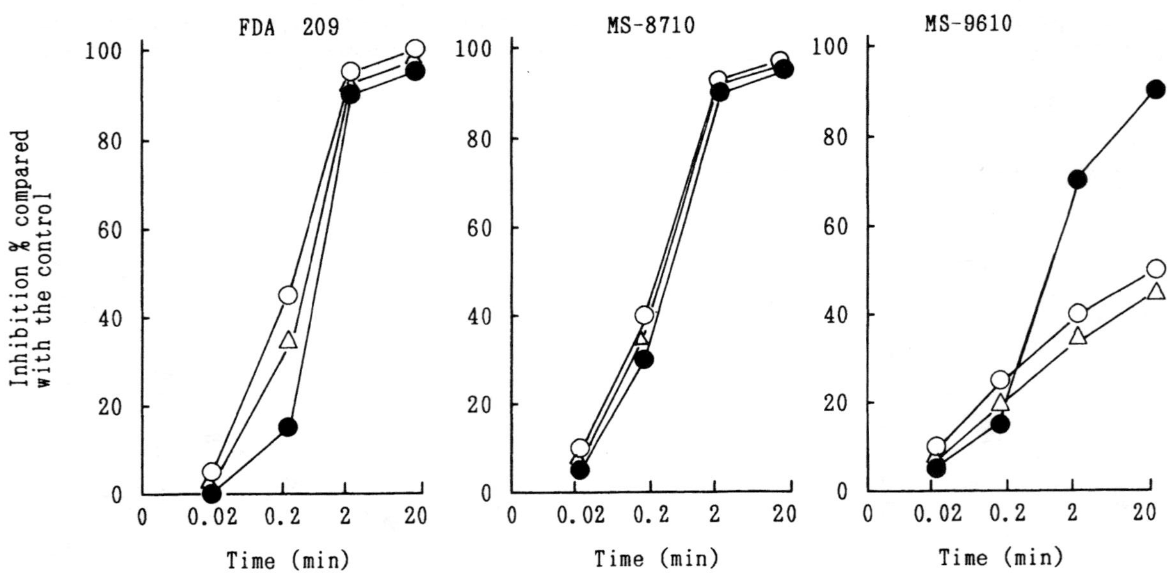

Figure 44. Effect of 3-acetyl-4″isovaleryltylosin, 3-acetyltylosin, and tylosin on poly(U)-directed polyphenylalanine synthesis with 70S-ribosomes from MS-8710, MS-9610 (resistant strains), and FDA 209P (a sensitive strain) of *Staphylococcus aureus*. ○: Tylosin; ●: 3-acetyl-4″isovaleryltylosin; △: 3-acetyltylosin.

was less than 10% at most. However, the yield was rapidly increased to about 90% by the combination of a strain-improvement study, that is, multiple NTG- and UV-mutation cycles, and the optimization of conversion conditions such as feed rate of substrates, pH, dissolved oxygen, and temperature. Furthermore, a mutant strain that could use L-leucine as a isovaleryl donor in place of isovaleryl CoA was successfully isolated, resulting in a marked decrease of raw material cost for production of 3-acetyl-4″-isovaleryltylosin.

Figure 45 shows a typical time course of conversion from tylosin to 3-acetyl-4″-isovaleryltylosin in a 20-liter jar fermentor (working volume 15 liters). After a 42-h batch cultivation of *S. thermotolerans* strain No. YN-1784, a mixed solution of tylosin and L-leucine was continuously supplied to the broth. A glucose solution was fed

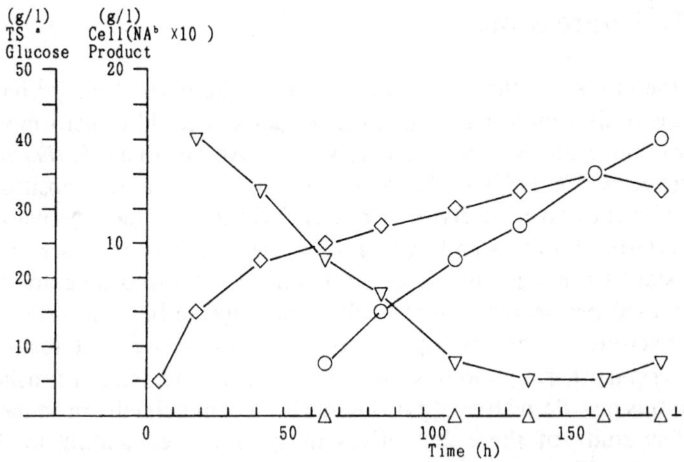

Figure 45. Time course of microbial transformation by *S. thermotolerans* strain no. YN-1784 from tylosin to 3-acetyl-4″-isovaleryltylosin. Cultivation medium (% w/v): corn starch, 3.0; soybean flour, 4.0; yeast extract, 0.1; $MgSO_4 \cdot 7H_2O$, 1; KH_2PO_4, 0.05; sodium citrate, 0.1; dry yeast, 1.5. Cultivation was carried out at 32°C and pH was controlled at 7.5. A mixture solution of tylosin and L-leucine (isovaleryl donor) and glucose solution was fed separately to the cultured-broth after a 42-h batch culture. ○: 3-Acetyl-4′-isovaleryltylosin; ▽: total sugar; △: glucose; ◇: Cell (expressed as nucleic acid). [a]total sugar; [b]nucleic acid.

separately at an appropriate flow rate so that the concentration of glucose in the broth was kept nearly at zero, thereby suppressing the overgrowth of the microorganism. As clearly seen from the figure, the conversion from tylosin to 3-acetyl-4″-isovaleryltylosin proceeded at a linear rate during the feeding of tylosin and L-leucine. Finally, 3-acetyl-4″-isovaleryltylosin was converted from tylosin at a yield of 88% in this run.

The process for recovery and purification of 3-acetyl-4″-isovaleryltylosin from the cultured broth has also been intensively studied. At present, the product is recovered and purified with the conventional solvent-extraction process followed by the precipitation process. The quality of the final product completely meets the requirements of customers and standards sets.

4.7. Future Scope

In the course of the screening program, Okamoto et al.[25,26] have successfully found that some microorganisms could transform the known macrolides to new acyl derivatives. Among them, *S. thermotorelans* ATCC 11416 acylated in particular at the 3 and 4″ positions of tylosin to give 3-acetyl-4″-isovaleryltylosin. This new derivative of tylosin showed significant antibacterial activity to macrolide-resistant strains due to increased affinity to 70S ribosome and/or increased permeability to the cell of pathogenic bacteria resistant to macrolides. The development of 3-acetyl-4″-isovaleryltylosin as verterinary therapy and feed additives has been successfully finished and this new derivative of tylosin has been commercially produced.

The study of these derivatives of tylosin is expanding in the following directions:

1. *Direct production of 3-acetyl-4″-isovaleryltylosin.* The tylosin-producing microorganisms strains produce only tylosin of the type in which the OH group is attached to the 3 and 4″-positions of tylosin. Consequently, the present production process for 3-acetyl-4″-isovaleryltylosin consists of the fermentative production of tylosin by *S. fradiae* and the subsequent bioconversion of tylosin to 3-acetyl-4″-isovaleryltylosin by *S. thermotolerans*. In order to simplify this two-step process. Arisawa

et al.[116] are studying the breed of a microorganism strain capable of making the 3- and 4″-acylation product of tylosin by genetic engineering. They have already succeeded in cloning genome gene of tylosin-4″-acyltransferase of *S. thermotolerans* into *S. fradiae* via a newly constructed plasmid. This transformant strain was already confirmed to produce 4″-isovaleryltylosin without addition of tylosin.[117] Further investigation is intensively in progress to introduce another acyltransferase, that is, 3-acyltransferase into *S. fradiae.*

2. *Elucidation of the structure-activity relationship.* Tsuchiya et al.[118] have studied the relationship between chemical structures and biological activities of newly chemically synthesized 49 acyl derivatives of macrolides and obtained the following results:
   (a) The introduction of a branched-chain aliphatic acyl group such as the 2-methoxyisovaleryl or the 4-methoxyvaleryl group increases the antibacterial activity of the derivatives.
   (b) Further improved activity was obtained by the introduction of aromatic group such as phenylthioacetyl, phenysulfonylacetyl, 4-nitrophenylacetyl, 4-nitrophenylsulfonyl, and phenylethanesulfonyl groups. These derivatives had also improved antimycoplasma activity.
   (c) The introduction of previously mentioned groups into 4″ position was confirmed to increase the uptake by resistant strains.

Kiyoshima et al.[119] have also studied the chemical structure-activity relationships by using newly isolated analogues produced by a block mutant strain of *S. fradiae*, a producer of tylosin. They found that the third sugar moiety of tylosin is very important for expression of the antimicrobial activity.

These findings will be advantageously reflected in future developments of tylosin derivatives that are effective *in vitro* and *in vivo* against macrolide-resistant isolates.

References

1. T. Nara, I. Kawamoto, R. Okachi, and T. Oka. *Jpn. J. Antibiot.*, **30**, S-174 (1977).
2. T. Nara, M. Yamamoto, I. Kawamoto, K. Takayama, R. Okaci, and S. Sato. *J. Antibiot.*, **30**, 533 (1977).
3. R. Okachi, S. Takasawa, T. Sato, S. Sato, M. Yamamoto, I. Kawamoto, and T. Nara. *J. Antibiot.*, **30**, 541 (1977).
4. S. Sato, T. Iida, R. Okachi, K. Shirahata, and T. Nara. *J. Antibiot.*, **30**, 1025 (1977).
5. M. Ohjoshi, J. Isigami. *Chemotherapy* (Japanese), **31**, 976 (1983).
6. R.S. Egan, R.S. Stanaszek, M. Cirovic, S.L. Mueller, J. Tadanier, J.R. Martin, P. Collum, A.W. Goldstein, R.L. DeVault, A.C. Sinelair, E.E. Fager, and L.A. Mitcher. *J. Antibiot.*, **30**, 552 (1977).
7. K. Shirahata, G. Shimura, S. Taksawa, T. Iida, and K. Takahashi. *Am. Chem. Soc., Symp. Ser.*, No. 125, 1980, p. 309.
8. Y. Ohasi, K. Yamashita, H. Kawabe, S. Mitsuhasi, K. Sato, and N. Nakamura. *Chemotherapy* (Japanese), **29**, (S-2), 1 (1981).
9. M. Goto, S. Mitsuhasi, M. Ogawa. A. Tuji, Y. Kaneko, and S. Kuwahara. *Chemotherapy* (Japanese), **29**, (S-2), 30 (1981).
10. M. Akiyoshi, S. Yano, H. Nakata, T. Nara, S. Nishikawa, T. Hara, and T. Ohkuro. *Chemotherapy* (Japanese), **29**, (S-2), 99 (1981).
11. M. Akiyoshi. In *Aminoglyoside Antibiotics* (Japanese), ed. by Y. Ueda, Nankodo, Tokyo, 1985, p. 123.
12. Y. Ueda, A. Saito, J. Simada, M. Omori, T. Shiba, T. Yamaji, H. Ihara, T. Kitajo, M. Kaji, and T. Miyahara. *Chemotherapy* (Japanese), **29**, (S-2), 99 (1981).
13. A. Saito and T. Yamaji. In *Aminoglyoside Antibiotics* (Japanese), ed. by Y. Ueda, Nankodo, Tokyo, 1985, p. 109.
14. T. Nara, M. Yamamoto, I. Kawamoto, K. Takayama, R. Okachi, S. Takasawa, T. Sato, and S. Sato. *J. Antibiot.*, **30**, 533–540 (1977).
15. K. Ohba, T. Shomura, T. Tsuruoka, M. Kojima, S. Inoue, and T. Ito. *Jpn. Kokai*, 18600, (Feb 21, 1981).
16. Y. Okami, K. Hotta, M. Yoshida, D. Ikeda, S. Kondo, and H. Umezawa. *J. Antibiot.*, **32**, 964–966 (1979).
17. T. Deushi, A. Iwasaki, K. Kamiya, T. Mizoguchi, M. Nakayama, H. Itoh, and T. Mori. *J. Antibiot.*, **32**, 1061–1065 (1979).
18. A. Iwasaki, H. Ito, and T. Mori. *J. Antibiot.*, **32**, 180–186 (1979).
19. T. Shomura, M. Kojima, J. Yoshida, M. Ito, S. Amano, K. Totsu-

gawa, T. Niwa, S. Inoue, T. Ito, and T. Niida. *J. Antibiot.*, **33**, 924-930 (1980).

20. N. Hirayama, K. Shirahata, Y. Ohashi, and Y. Sasada. *Molec. Pharmacol.*, **23**, 127-132 (1983).
21. S. Itoh, Y. Odakura, H. Kase, S. Sato, K. Takahashi, T. Iida, K. Shirahata, and K. Nakayama. *J. Antibiot.*, **37**, 1664-1669 (1984).
22. Y. Odakura, H. Kase, S. Itoh, S. Satoh, S. Takasawa, K. Takahashi, and K. Shirahata. *J. Antibiot.*, **37**, 1670-1680 (1984).
23. J.R.D. McCormick, U. Hirsch, N.O. Sjolander, and P.A. Doerschuck. *J. Am. Chem. Soc.*, **82**, 5006-5009 (1960).
24. K. Shirahata, H. Kase, S. Kitamura, and T. Iida. *J. Antibiot.*, **35**, 520-523 (1982).
25. K.L. Rinehart, Jr., and R.M. Stroshane. *J. Antibiot.*, **29**, 319-353 (1976).
26. W.H. Horner. *J. Biol. Chem.*, **239**, 578-581 (1964).
27. M.S. Walker, and J.B. Walker. *J. Biol. Chem.*, **241**, 1262-1269 (1966).
28. J.B. Walker, and M.S. Walker. *Biochemistry*, **6**, 3821-3829 (1967).
29. H. Kase, Y. Odakura, and K. Nakayama. *J. Antibiot.*, **35**, 1-9 (1982).
30. T. Futami, K. Takeda, A. Kinumaki, Y. Ito, and T. Okuda. *J. Antibiot.*, **32**, 891-899 (1979).
31. S.J. Daum, D. Rosi, and W.A. Goss, *J. Antibiot.*, **30**, 98-105 (1977).
32. H. Kase, S. Kitamura, and K. Nakayama. *J. Antibiot.*, **35**, 385-390 (1982).
33. K. Hotta, S. Morioka, H. Tohyama, and Y. Okami. In *Trends in Actinomycetology in Japan*, ed. by Y. Koyama, Society for Actinomycetes, Japan, 1989, pp. 61-64.
34. K.F. Chater, D.A. Hopwood, T. Kieser, and C.J. Thompson. *Curr. Top. Microbiol. Immunol.*, **97**, 69-95 (1982).
35. T. Dairi, and M. Hasegawa. *J. Antibiot.*, **42**, 934-943 (1989).
36. T. Ohta, E. Hashimoto, T. Dairi, and M. Hasegawa. In *Trends in Actinomycetology in Japan*, ed. by Y. Koyama, Society for Actinomycetes, Japan, 1989, pp. 101-102.
37. E. Higashide, T. Fugono, T. Hatano, and M. Shibata. *J. Antibiot.*, **24**, 1 (1971).
38. S. Harada, T. Kishi, and K. Mizuno. *J. Antibiot.*, **24**, 13 (1971).
39. S. Harada, E. Higashide, T. Fugono, and T. Kishi. *Tetrahedron Lett.*, **27**, 2239 (1969).
40. S. Harada, and T. Kishi, *Chem. Pharm. Bull.*, 22, 99 (1974).
41. K. Tsuchiya, T. Yamazaki, Y. Takeuchi, and T. Oishi, *J. Antibiot.*, **24**, 29 (1971).

42. K. Kamiya, S. Harada, K. Wada, M. Nishikawa, and T. Kishi. *Tetrahedron Lett.*, **27**, 2245 (1969).
43. T. Fugono, E. Higashide, T. Suzuki, H. Yamamoto, S. Harada, and T. Kishi. *Experientia*, **26**, 26 (1970).
44. T. Yamazaki, N. Narukawa, I. Suenaga, and K. Takeda. *Proc. 9th Congr. Int. Pig Vet. Soc.*, Spain (1986), p. 178.
45. K. Tsuchiya, T. Yamazaki, Y. Takeuchi, and T. Oishi, *J. Antibiot.*, **24**, 29 (1971).
46. T. Hayashi, I. Suenaga, N. Narukawa, and T. Yamazaki, *Antimicrob. Agents Chemother.*, **32**, 458 (1988).
47. H. Ishikawa, S. Shimaie, T. Ohsawa, Y. Wada, J. Noshima, and K. Tokunaga. *Annu. Meet. Jpn. Vet. Med. Assoc.*, Tokyo (1985), p. 262.
48. J. Okada, and S. Kondo, *J. Assoc. Offic. Anal. Chem.*, **70**, 818 (1987).
49. J. Okada, S. Yamamoto, H. Yamamoto, and S. Kondo, *J. Agr. Food Chem.*, **36**, 104 (1988).
50. S. Harada, J. Okada, M. Takeda, and T. Yamazaki. *J. Antibiot.*, **38**, 877 (1985).
51. H. Sawada, T. Suzuki, S. Akiyama, and Y. Nakao. *Appl. Microbiol. Biotechnol.*, **26**, 522 (1987).
52. K. Hatano, S. Harada, and T. Kishi. *J. Antibiot.*, **28**, 15 (1975).
53. H. Sawada, T. Suzuki, S. Akiyama, and Y. Nakao. *Appl. Microbiol. Biotechnol.*, **32**, 556 (1990).
54. T. Korpela, and J. Himanen. *J. Chromatogr.*, **290**, 351 (1984).
55. M. Rosenberg, D. Gutnick, and E. Rosenberg. *FEMS Microbiol. Lett.*, **9**, 29 (1980).
56. S. Harada, T. Kishi, and K. Mizuno. *J. Antibiot.*, **24**, 13 (1971).
57. H. Sawada, T. Suzuki, S. Akiyama, and Y. Nakao. *Appl. Microbiol. Biotechnol.*, **26**, 522 (1987).
58. H. Sawada, T. Suzuki, S. Akiyama, and Y. Nakao. *Appl. Microbiol. Biotechnol.*, **32**, 556 (1990).
59. M. Uramoto, N. Otake, L. Cary, and M. Tanabe. *J. Amer. Chem. Soc.*, **100**, 3616 (1978).
60. K. Hatano, S. Harada, and T. Kishi. *J. Antibiot.*, **28**, 15 (1975).
61. C. Huggins, and J. Lapides. *J. Biol. Chem.*, **170**, 467 (1947).
62. A. Matsushima, H. Nishimura, Y. Ashihara, Y. Yokota, and Y. Inada. *Chem. Lett.*, **1980**, 773 (1980).
63. A. Asuchowski, T.V.Es, N.C. Palczuk, and F.F. Davis. *J. Biol. Chem.*, **252**, 3578 (1977).
64. H. Lineweaver, and D. Burk. *J. Amer. Chem. Soc.*, **56**, 659 (1934).

65. K. Tachibana and K. Kaneko. *J. Pesticide Sci.*, **11**, 297–304 (1986).
66. H. Takebe, S. Imai, H. Ogawa, A. Satoh and H. Tanaka. *J. Ferment. Technol.*, **67**, 226–232 (1989).
67. H. Takebe, M. Matsunaga, O. Hiruta, A. Satoh and H. Tanaka. *J. Ferment. Technol.* (in press).
68. H. Takebe, M. Matsunaga, O. Hiruta, A. Satoh and H. Tanaka. *J. Ferment. Technol.* (in press).
69. R.P. Elander, et al. *Folia Microbiol.*, **16**, 156–165 (1971).
70. S.A. Goulden and F.W. Chattaway. *J. Gen. Microbiol.*, **59**, 111–118 (1969).
71. J.D. Bu'Luck, D. Hamilton, M.A. Hulme, A.J. Powell, H.M. Smalley, D. Shepherd and G.N. Smith. *Can. J. Microbiol.*, **11**, 765–778 (1965).
72. A. Jones and D.W.S. Westlake. *Can. J. Microbiol.*, **20**, 1599–1611 (1974).
73. M. Gallo and E. Katz. *J. Bact.*, **109**, 659–667 (1972).
74. H. Seto, S. Imai, T. Tsuruoka, A. Satoh, M. Kojima, S. Inoue, T. Sasaki and N. Otake. *J. Antibiotics.*, **35**, 1719–1721 (1982).
75. Z. Hostalek, M. Tinterova, V. Jechova, M. Blumauerova, J. Suchy and Z. Vanek. *Biotech. Bioeng.*, **11**, 539–548 (1969).
76. H. Okada, I. Kameyama, S. Okumura and T. Tsunoda. *J. Gen. Appl. Microbiol.*, **7**, 177 (1961).
77. C.M. Stevens, E.P. Abraham, F.C. Huang and C.J. Sih. *Federation Proc.*, **34**, 625 (1975).
78. E. D. Eeinberg. *Perspect. Biol. Med.*, **5**, 432 (1962).
79. Z. Hostalek et al. Academia Prague, 1973, p. 353.
80. H. Shibai, A. Ishizaki and Y. Hirose. *Agr. Biol. Chem.*, **37**, 2083–2088 (1973).
81. K. Akashi, H. Shibai and Y. Hirose. *Hakko to Kogyo* (Japan)., **37**, 321–330 (1979).
82. K. Akashi, H. Shibai and Y. Hirose. *Hakko to Kogyo* (Japan)., **40**, 518–526 (1982).
83. H. Seto, T. Sasaki, S. Imai, T. Tsuruoka, H. Ogaa, A. Satoh, S. Inoue, T. Niida and N. Otake. *J. Antibiot.*, **36**, 96–98 (1983).
84. H. Seto, S. Imai, T. Tsuruoka, H. Ogawa, A. Satoh, M. Kojima, T. Sasaki, and N. Otake. *Biochem. Biophy. Res. Commun.*, **111**, 1008–1014 (1983).
85. H. Ogawa, S. Imai, T. Shimizu, A. Satoh, and M. Kojima. *J. Antibiot.*, **36**, 1040–1044 (1983).
86. I. Nakayama. In *Macrolide Antibiotics*, ed. S. Omura, London, 1984, pp. 261–300.

87. R.C. Wilson. In *Macrolide Antibiotics*, ed. S. Omura, London, 1984, pp. 301–347.
88. S. Mitsuhashi and M. Inoue. In *Antimicrobial Drug Resistance*, ed. L.E. Bryan, London, 1984, pp. 279–291.
89. M. Nakae, T. Murai, Y. Kaneko, and S. Mitsuhashi. *Antimicrob. Agents Chemother.*, **12**, 427–428 (1977).
90. Y. Miyamoto, T. Takizawa, and A. Matsushima. *Antimicrob. Agents Chemother.*, **13**, 399–404 (1978).
91. S. Omura. In *Antibiotics* (Japanese), ed. H. Yonehara, Tokyo, 1981, pp. 187–228.
92. S. Omura and A. Nakagawa. *J. Antibiot.*, **28**, 401–433 (1975).
93. N.P. Leon and V. Jean. U.S. Patent, 2943024 (1978).
94. T. Osono, K. Yano, K. Murakami, F. Miyamoto, and K. Miriyama. *Jpn. J. Antibiot.*, **26**, 115–120 (1973).
95. S. Harada, M. Muroi, M. Kondo, K. Tsuchiya, T. Matsuzawa, T. Fugono, T. Kishi, and J. Ueyanagi. *Antimicrob. Agents Chemoth.*, **4**, 140–148 (1973).
96. S. Inoue, T. Shomura, T. Tsuruoka, S. Omoto, T. Niida, and K. Umemura. *Chem. Pharm. Bull.*, **20**, 2366–2371 (1972).
97. L.I. Feldman, I.K. Dill, C.E. Holmilund, H.A. Whaley, E.L. Patterson, and N. Bohonos. *Antimircrob. Agent. Chemoth.*, 54–57 (1964).
98. K. Nakayama and S. Igarashi. *J. Antibiot.*, **27**, 605–609 (1974).
99. M. Shibata, M. Uyeda, and S. Mori. *J. Antibiot.*, **28**, 434–441 (1975).
100. K. Nakayama, M. Izawa, M. Muroi, T. Kishi, M. Uchida, and S. Igarashi. *J. Antibiot.*, **27**, 425–432 (1974).
101. K. Nakayama, T. Kishi, and S. Igarashi. *J. Antibiot.*, **27**, 487–488 (1974).
102. M. Uyeda, S. Mori, M. Morita, and T. Ogata, M. Mori, and M. Shibata. *J. Antibiot.*, **30**, 1130–1131 (1974).
103. A. Tachibana, M. Shibata, F. Kumagai, K. Yano, and T. Osono. *Jpn. J. Antibiot.*, **26**, 122–129 (1973).
104. K. Nakayama, T. Kishi, and S. Igarashi. *J. Antibiot.*, **27**, 433–441 (1973).
105. R.J. Adamski, H. Heymann, S.G. Geftic, and S.S. Barkulls. *J. Med. Chem.*, **9**, 933–934 (1966).
106. H. Takahira. *Jpn. J. Antibiot.*, **23**, 424–428 (1970).
107. S. Omura, M. Katagiri, I. Umezawa, K. Komiyama, T. Maekawa, K. Sekikawa, A. Matsumae, K. Sekiyama, A. Matsumae, and T. Hatta. *J. Antibiot.*, **21**, 532–538 (1968).
108. R.B. Morin, M. German, P.L. Hamill, and P.V. Demarco. *Tetrahedron Lett.*, 4737–4740 (1970).

109. T. Furumai, Y. Seki, A. Takeda, A. Kinumaki, and M. Suzuki. *J. Antibiot.*, **26**, 708–710 (1973).
110. R. Okamoto, H. Nomura, M. Tsuchiya, H. Tsunekawa, T. Fukumoto, T. Inui, T. Sawa, T. Takeuchi, and H. Umezawa. *J. Antibiot.*, **32**, 542–544 (1979).
111. R. Okamoto, T. Fukumoto, K. Imafuku, T. Okubo, K. Kiyoshima, A. Takamatsu, and T. Takeuchi. *J. Ferment. Technol.*, **57**, 519–528 (1979).
112. R. Okamoto, M. Tsuchiya, H. Nomura, H. Iguchi, K. Kiyoshima, S. Hori, T. Inui, T. Sawa, T. Takeuchi, and H. Umezawa. *J. Antibiot.*, **33**, 1309–1315 (1980).
113. R. Okamoto, T. Fukumoto, H. Nomura, K. Kiyoshima, K. Nakamura, A. Takamatsu, H. Naganawa, T. Takeuchi, and H. Umezawa. *J. Antibiot.*, **33**, 1300–1305 (1980).
114. M. Tsuchiya, K. Suzukake, M. Hori, T. Sawa, T. Takeuchi, H. Umezawa, R. Okamoto, H. Nomura, H. Tsunekawa, and T. Inui. *J. Antibiot.*, **34**, 305–312 (1980).
115. M. Tsuchiya, T. Sawa, T. Takeuchi, H. Umezawa, and R. Okamoto. *J. Antibiot.*, **35**, 673–679 (1981).
116. A. Arisawa, N. Kawamura, K. Okamura, H. Tone, and R. Okamoto. In *Abstr. Papers, Annual Meeting of Agric. Chem. Soc. Japan*, Niigata, 1989, p. 146.
117. N. Kawamura, A. Arisawa, I. Kojima, K. Okamura, H. Tone, and R. Okamoto. In *Abstr. Papers, Annual Meeting of Agric. Chem. Soc. Japan*, Niigata, 1989, p. 146.
118. M. Tsuchiya, M. Hamada, T. Takeuchi, H. Umezawa, K. Yamamoto, H. Tanaka, K. Kiyoshima, S. Mori, and R. Okamoto. *J. Antibiot.*, **35**, 661–672 (1981).
119. K. Kiyoshima, K. Takada, M. Yamamoto, K. Kubo, R. Okamoto, Y. Fukagawa, and T. Ishikura. *J. Antibiot.*, **40**, 1123–1130 (1987).

Subject Index

Japanese Technology Reviews

Section A: Electronics

Volume 1, Number 1
MMIC—MONOLITHIC MICROWAVE INTEGRATED CIRCUITS
Yasuo Mitsui

Volume 1, Number 2
BULK CRYSTAL GROWTH TECHNOLOGY
Shin-ichi Akai, Keiichiro Fujita, Masamichi Yokogawa, Mikio Morioka and Kazuhisa Matsumoto

Volume 2, Number 1
SEMICONDUCTOR HETEROSTRUCTURE DEVICES
Masayuki Abe and Naoki Yokoyama

Volume 2, Number 2
DEVELOPMENT OF OPTICAL FIBERS IN JAPAN
Hiroshi Murata

Volume 3, Number 1
HIGH-PERFORMANCE BiCMOS TECHNOLOGY AND ITS APPLICATIONS TO VLSIs
Ikuro Masuda and Hideo Maejima

Volume 3, Number 2
SEMICONDUCTOR DEVICES FOR ELECTRONIC TUNERS
Seiichi Watanabe

Section B: Computers and Communications

Volume 1, Number 1
MACHINE VISION: A PRACTICAL TECHNOLOGY FOR ADVANCED IMAGE PROCESSING
Masakazu Ejiri

Volume 1, Number 2
CRYPTOGRAPHY AND SECURITY
Edited by Shigeo Tsujii

Volume 2, Number 1
VLSI NEURAL NETWORK SYSTEMS
Yuzo Hirai

Section C: New Materials

Section D: Manufacturing Engineering

Volume 1, Number 1
AUTOMOBILE ELECTRONICS
Shoichi Washino

Volume 1, Number 2
STEEL INDUSTRY I: MANUFACTURING SYSTEM
Tadao Kawaguchi and Kenji Sugiyama

Volume 2, Number 1
STEEL INDUSTRY II: CONTROL SYSTEM
Tadao Kawaguchi and Takatsugu Ueyama

Volume 2, Number 2
NETWORKING IN JAPANESE FACTORY AUTOMATION
Koichi Kishimoto, Hiroshi Tanaka, Yoshio Sashida and Yasuhisa Shiobara

Section E: Biotechnology

Volume 1, Number 1
PRODUCTION OF NUCLEOTIDES AND NUCLEOSIDES BY FERMENTATION
Sadao Teshiba and Akira Furuya

Volume 1, Number 2
RECENT PROGRESS IN MICROBIAL PRODUCTION OF AMINO ACIDS
Hitoshi Enei, Kenzo Yokozeki and Kunihiko Akashi

Volume 2, Number 1
RECENT ADVANCES IN JAPANESE BREWING TECHNOLOGY
Takashi Inoue, Jun-ichi Tanaka and Shunsuke Mitsui

Volume 2, Number 2
ANTIBIOTICS I: β-LACTAMS AND OTHER ANTIMICROBIAL AGENTS
Isao Kawamoto and Masao Miyauchi

Volume 3, Number 1
ANTIBIOTICS II: ANTIBIOTICS BY FERMENTATION
Sadao Teshiba, Mamoru Hasegawa, Takashi Suzuki, Yoshiharu Tsubota, Hidehi Takebe, Hideo Tanaka, Mitsuyasu Okabe and Rokuro Okamoto

Volume 3, Number 2
OLIGOSACCHARIDES: PRODUCTION, PROPERTIES, AND APPLICATIONS
Edited by Teruo Nakakuki

For Product Safety Concerns and Information please contact our EU representative GPSR@taylorandfrancis.com Taylor & Francis Verlag GmbH, Kaufingerstraße 24, 80331 München, Germany

Batch number: 10725869

Printed by Printforce, the Netherlands